ROBOTICS

Also by John F. Young.

Cybernetics
Applied Electronics
Information Theory
Cybernetic Engineering

ROBOTICS

JOHN F. YOUNG

M.Sc., B.Sc., C.G.I.A., M.I.E.E., M.I.E.R.E.
Cybernetics Laboratory, University of Aston

A HALSTED PRESS BOOK

JOHN WILEY & SONS
New York — Toronto

Published in the U.S.A. and
Canada by Halsted Press, a
Division of John Wiley & Sons, Inc.,
New York.

Library of Congress Cataloging in Publication Data

Young, John Frederick.
  Robotics.

  "A Halsted Press book."
  Includes bibliographies.
  1. Automata.   I. Title.
  TJ211.Y68   629.8'92   72–6308
  ISBN 0–470–97990–9

First published in 1973

© John F. Young, 1973

All rights reserved by
Butterworth & Co. (Publishers) Ltd., London

Printed in England by Redwood Press Limited, Trowbridge, Wiltshire

Preface

The general-purpose robot is even now rapidly emerging from the laboratory. There are already well over one hundred manufacturers of such devices in the world, and robots are quietly being introduced into factories in all industrial countries. In many cases these robots are directly replacing human workers.

At the present time over 7% of the world's manufacturers of robots are Japanese companies and the Japanese Government has decided to spend some 100 000 000 dollars on robotics research and development over the next few years. There are many reasons for the growth of this field, but perhaps the outstanding reasons, particularly in the Western World, are efficiency and consistency. The robot never tires or loses interest in its task. It can go on performing twenty-four hours each day if necessary, with only an occasional weekly or monthly maintenance session, and will carry out its allotted task consistently without boredom or fatigue. Some robots have merely to be taken through their paces once by a human teacher, on a task which might occupy several minutes of complex movements, and they can then repeatedly perform this complex task without supervision day in and day out. The human worker cannot compete. It is an unfortunate fact that today, even in countries which are technologically very far advanced, many people still have to earn their living undertaking repetitive tasks in which there is no room for individual creativity.

Now at last we can see an end to these conditions. The efficient and flexible machine robots can release humans to become creative creatures, but this will undoubtedly involve a great deal of economic upheaval. If the robots are merely allowed to replace people who then become redundant, the resulting unemployment will simply cause the economy to collapse. The first requirement for any thriving economy is a ready supply of prosperous consumers.

One has to consider what might happen if robots, produced in large quantities and therefore at low cost, are introduced into countries where there is already an excessive pool of unemployment. It is hoped that the very extent of the progress which has been made already in research and development in the field of robotics, as described in the present book, will make it clear that we should be considering such problems very urgently at the present time.

There is another aspect of robotics research and development which should be seriously considered. Various organisations are known to be developing robot devices capable of carrying out, unsupervised, various domestic tasks. Such robots will be extensions of our present-day domestic appliances, with the important difference that they will be mobile devices which are capable of learning. If these robots can be produced at a reasonably low cost, and all present indications are that this is possible, then clearly there will be a big demand for such devices. Thus while low-cost robots will be welcomed for domestic use their appearance would mean that the cost of robot devices had fallen to the point where wide-scale industrial utilisation was inevitable. Such is the nature of the general-purpose robot that further human redundancy would then become unavoidable.

This book covers all aspects of robotics with the exception of the controlling 'brain'; this has been dealt with in my book *Cybernetic Engineering*. The characteristics of human sense organs are reviewed as an introduction to robot methods of perception and control. The advantages and disadvantages of the different possible methods of robot actuation, i.e. pneumatic, electrical, etc., are compared and practical industrial robots such as the Unimate and the Versatran, together with recently introduced devices and undersea manipulators, are described in detail. The information covers the theory and practice of robotics and will be of interest to all those working in or entering this new field. It is hoped that the extensive bibliographies given at the end of each chapter will enable the reader to follow up in more detail any particular aspect of the subject.

JOHN F. YOUNG

Contents

1 Robots   1
Introduction—The form of the robot—Man's fear of robots—The laws of robotics—Thring's view of the robot—The domestic robot—Everyday robots—Robot passenger-carrying vehicles—Copying robots—Robot economics—Robots and redundancy—Robots for test and inspection—Robots for poor environments—Robot senses—References.

2 The Senses of the Robot   22
Human sensors—Response of receptor nerve cells—Human neuron numbers—Reflex action in the robot—Enhancement of contrast—Gas and moisture detection—Temperature regulation—Temperature control using thermistors—Force feedback sensors—Position feedback—Magnetic pulse generation—Resistive strain gauges—Velocity and acceleration feedback—References.

3 The Muscles of the Robot   42
The dynamic performance of muscle—Human transfer function—Performance requirements of hand and arm—The lifting action in the robot—Introduction of backlash—References.

4 Electrical Actuation   57
Practical requirements for actuators—General comparison of storage methods—Electrical actuation—The noise of electrical drives—Solenoids as actuators—Torque motors and rotary solenoids—Step-servo motors—Cooling—References.

5 Power for Actuation   72
Battery economy—Clutches and brakes—Pneumatic actuation—Hydraulic actuation—Batteries for electrical supply—Recent developments—Battery maintenance—Rapid charging of storage cells—References.

6 Robot Stability   91
Introduction—Simplified feedback stabilisation—Extension of simplified stabilisation—Electronic control of series motors—Field control of motors—Simple mobile machines—References.

7 Robot Mobility 106
Introduction—Robots for vehicle driving—The Robotug system—Robot urban transit vehicles—Safety with the Robotug—Automatic farm tractors—Flying robots—Space robots—Mobile manipulators—The Mascot mobile manipulator—Near field control—The rivet machine— Mobile machines controlled by a large computer—Legs, wheels or tracks—Lunokhod—References.

8 Robot Limbs 139
Practical robot arms and hands—Remotely controlled manipulators—Powered manipulators—Prosthetic hands and arms—Exoskeletons—Balance sensors—Other methods—The robot storekeeper—Human walking—Flexible knee joint—Ankle joints—Large quadruped walking machine—Other anthropomorphous machines—References.

9 Practical Robots 163
Copying robots for industrial use—Mechanical hands on machine tools—The Planobot—The Unimate—Versatran devices—Miscellaneous robot devices—Comparative economics of robots and humans—Problems in special environments—Undersea manipulators—References.

10 Human Vision 187
Development of human vision—Information capacity of the human eye—Other features of the eyes—Binocular vision—Colour sensitivity—Optical illusions—Cleanliness of robot eyes—References.

11 Character Recognition 200
Visual character recognition—Perceptrons—The Shadow mask technique—Other inhibitory forms—Results with the Mk 17 Program—Incoherent Fourier transformation—A scanning artificial retina—Simple pattern-matchnig gates—References.

12 Robot Vision 212
Minimal character recognition—Extension of the minimal method—Edge detection at the retina—Contrast enhancement in a scanning system—Outline enhancement by superposition—A counting retina—An edge detecting retina—Future possibilities for retinas—Edge detection in a computer scanning system—References.

13 Perception of Movement 228
Eye movement—Eye tremors—Fibre optics—Fibre optic eye—The eye following a moving object—Automatic focusing of the robot eye—Persistence of vision—Miniature radars and Doppler effect—References.

14 Human Hearing 242
The robot ear—Work on speech recognition—The nature of speech—Information content of human speech—Limitation of frequency bandwidth—The single equivalent formant—The structure of spoken words—Miscellaneous points—Audible illusions—Binaural hearing—Microphones—References.

15 Robot Hearing and Speech   257
Machines to recognise human language—Vocoder using passive filters—Active filters for Vocoder—Practical robot Vocoder using active filters—Mechanical filters in a Vocoder—An approach to numeral recognition—Robot speech—Detection of self-produced signals—Inductive and capacitive proximity detectors—References.

16 Robot Reliability   274
Finite life of robots—Logical selection—The effectiveness of a robot—Mean time between failures—Over-all failure pattern—Redundant parts——Repair time—References.

17 The Future of the Robot   291

Index   293

1

Robots

Introduction

The word 'cybernetics' seems to mean many different things to different people[1]. In the minds of the general public, cybernetics appears to be associated with 'computers, or robots, or something'. How valid is such a view?

Certainly one eventual aim of cybernetic engineering and of Robotics is to produce robots of various types[55]. The writer's companion book *Cybernetic Engineering* deals with work on the 'brain' of the robot, while the present book deals with all other aspects. In particular, it is hoped that humanoid robots will be built some day. However, in addition to the humanoid, there are many other forms of robot device which can be envisaged. Indeed, there are many devices even now which have some of the characteristics of the robot and which are meeting a real need in the human environment.

There are many tasks, performed daily by human beings, which do not at all make use of the capabilities of the human[2, 37-39]. As a youngster in industry, the writer was engaged on some of these tasks. Not only is such work, simple, and repetitive in the extreme, suitable for performance by a machine, but a machine produces better results and fewer errors than does the mere erring human.

In the past it has been a simple question of economics which has decided in some cases that a human shall continue to do degrading repetitive or dirty work. In the coming Robot Economy the position will be changed. If general-purpose robots can be made economically (and all indications are that this is the case), then they will be used to replace humans at work.

The form of the robot

Ever since the days of Homer, the fictional robot has traditionally been a mobile device and it has also been somewhat humanoid in appearance in general, though often rather metallic. In practice, however, it is unlikely that robots will in fact be manufactured in anything like a true humanoid form, at least in the near future.

Asimov has suggested[3] that the human form of the fictional robot is 'merely an anthropocentric fetish'. As probably the best-known writer on fictional robots, Asimov should know. The operation of a thermostat would not improve if it were man-shaped, with metal hands which turned the heating appliance on and off. However, if a robot is required to be capable of carrying out all of man's functions, Asimov suggests that it is best shaped like a man, since the human environment is adapted to that shape.

This is of course only partially true. We all know of motor cars and domestic appliances in which certain defective parts could only possibly be replaced by that mythical two-foot-tall midget with three arms, each four feet long and tipped with tentacle-like fingers.

The day is not far off when, if there is a real requirement for a number of robot devices meeting such a specification, then they will be made.

From an aesthetic point of view, man would no doubt prefer his robots to be humanoid in appearance. This would make them easier to live with. The extreme form of such a robot, even having a human-like skin covering making it indistinguishable from a human, has been called an android.

Man's fear of robots

The confusion which has sometimes been caused fictionally by robots which are excessively humanoid in appearance would be very easy to avoid. It is interesting to note that, until very recently, in fiction the humanoid robot has in general been a mechanism to fear. Frankenstein created a monster[4]. Rossum's Universal Robots destroyed all human life, or perhaps one should say almost all human life[5]. Must the robot destroy its creator?

Only in the past 30 years has the notion of the robot as a useful, properly engineered, safe machine become popular. The robot now is to be a friend of man. Or is he? It is noteworthy that Asimov, who seems to have first introduced the notion of the friendly, useful robot, was also responsible for the introduction into fiction of the famous 'Laws of Robotics'[6].

The laws of robotics

The laws of robotics are:
1. A robot may not injure a human being or, through inaction, allow a human being to come to harm.
2. A robot must obey the orders given to it by human beings, except where such orders would conflict with the First Law.
3. A robot must protect its own existence, except where such protection would conflict with the First or Second Law.

These laws contain admitted ambiguities, and they appear to have been investigated and modified by various people in various ways. However, it is a remarkable fact that this original fictional formulation can still provide a useful design guide to the cybernetic engineer.

The obvious example of this is in the design of mobile robot machinery, where the outstanding requirement is that the machine must conform to the First Law. The Second Law is much easier to meet. Not only must a mobile robot be made as incapable as possible of injuring a human by its movements, but also a very high order of reliability is called for in the electronic brain of the device. Because of this, investigations are being carried out using small mobile machines of the 'tortoise' type which will be described later. Because of their small size, these are inherently incapable of harming a human, but they are certainly capable of failure. It is in an attempt to reduce the failure rate that we are investigating such machines, with some success.

A terrible vision haunts the mind of the cybernetic engineer. It is of one of his mobile housework robots going wrong and chasing a screaming housewife—or, worse still, a screaming child—around a house. More serious accidents happen in industry when our present-day robot devices fail, but some risk is accepted in industry. However, a single domestic accident of the sort described could jeopardise the whole future of cybernetic engineering and of robotics.

Thring's view of the robot

One of the best-known workers in the field of practical robotics is Professor Meredith Thring, of Queen Mary College. His laboratory has constructed many different forms of robot device and of robot limbs. Thring has suggested[2] that the minimal robot must have:

1. A hand and an arm.
2. Self-propulsion and self-steering.

3. Power systems and control systems for (1) and (2).
4. A limited computer with memory for instructions and for decision making.
5. Senses of touch, roughness, hardness, position, weight, thermal conductivity, temperature, proximity, shape, size, sight, colour, distance, smell, position of limbs and hearing.

However, Thring feels that there are fundamental limitations to the robot. He suggests that a robot can never perform any task more sophisticated or organised than those it has been instructed to carry out by a human. He also suggests that if the robot makes errors, these will be random and not take the form of improvement or be self-correcting.

The present writer would suggest here that some of the random errors, if they are truly random, must of necessity cause improvements, and this must be true however we define 'improvement'. In addition, we know now that it is quite possible to make self-correcting and adaptive control systems, provided only that the eventual aim or purpose of the system can be quite clearly defined.

Thring proposes a hypothetical Aesthetic Law which states that man-made machines must always remain qualitatively inferior to the men who made them, and he suggests that a robot is barred from, for example, composing a symphony. Yet it must be pointed out that digital computers are capable, simply because of their speed, of performing calculations which would be quite impossible for any single man to perform in his limited lifetime. And, given a few basic rules of composition, digital computers have been shown to be capable of the production of quite original music, never before heard by man.

Some of the problems which Thring can see as remaining to be solved before the useful domestic robot is achieved are:

1. Some pattern recognition is required, so that the robot can divide objects into a few tens of categories, each of which has to be treated differently, e.g. chairs, tables, books, cutlery, crockery and so on.
2. The hand must be controlled so that it can pick up and deal with the objects seen by the robot, as, for example, in folding a sheet or peeling a vegetable.
3. Slight variations in the instructions of the robot must produce changes of the programme of work. Examples are the number of places to be laid at the table, changing the sheets on a bed, coping with changes of the position of furniture, etc.
4. The detection of dirt, and the quantity of dirt, so that, for

example, dish washing proceeds until all plates are clean and all cooking vessels are scoured.
5. A delicacy of touch is required, so that china or glass is not broken while being handled.

The domestic robot

Thring has suggested that there are two possible solutions to the problem of elimination of household drudgery. The first is that proposed by some architects, who have suggested that a home should be constructed so that no dust or dirt could ever enter—in fact, that it should look like a hospital operating theatre. The second solution would be much more acceptable to the average untidy human. It is to have a mechanical version of the Victorian housemaid to do all of the routine clearing up and cleaning around the house.
 Thring envisages such a robot as having a body some 1m high, with a single arm coming from the top of the machine and having two sections, each 1m long, with a strong hand at the end. This arm would then be able to reach anywhere in a sphere of 2m radius, centred 1m above the ground, so that the hand could reach to a height of 3 m and down to the ground in a circle of nearly 2 m radius. The hand should be able to pick up an object weighting 40 kg, or perhaps in some cases only 10 kg at full radius.
 Senses of touch and of vision would be required on such a domestic robot. Thring envisages such a robot laying and clearing tables for meals; loading dish-washing machines; making beds; washing floors, baths, sinks, walls; dusting, sweeping and clearing away rubbish; washing, ironing and drying clothes; and preparing food. Since food preparation is a creative human activity, the robot would probably be restricted in the kitchen to such tasks as peeling potatoes.

Everyday robots

The robot is with us already. Robot devices have appeared quite innocently and we are so familiar with them that we no longer even notice their existence[24]. We drive along the road and we obey the signals given to us by the robot traffic signals which replace the policeman waving his arms. One thing that amuses foreigners when in this country is that the Englishman still obeys the signals given by these robots at two a.m., when there is little or no traffic on the roads. In the kitchen the housewife passes on some of her chores to her robot washing machine and to her robot dish-washer. We make a telephone

6   *Robots*

call to Europe, and the switching is completely handled by robots: by quite exceptional robots, which do their best to get the call through by some route, even if some part of an exchange has broken down.

The digital computer is a form of non-mobile robot which is now influencing all of our lives. In some cases programmes produced by a computer are being used to control the operation of machine tools in the factory[27, 28]. At present such programmes are rather inflexible and they cannot cope with unexpected breakdowns. However, the reliability of such equipment is being improved all the time.

In industry very long and complex automatic production lines are now in use. These are quite automatic, and they have to be very reliable, since the cost of a shut-down can be as much as £1000 each minute of down-time. However, such lines do break down on occasions.

The writer was fortunate in being able to observe one of the very infrequent shut-downs of the automatic production line in a factory producing car engine blocks. The line stopped, and men began to run in all directions. An electrician ran to an electronic control cubicle near to the writer. He glanced at the pattern of lights in the cubicle. Then he muttered to himself, 'Mechanical fault on number six!' He dashed away to 'head' number six on the production line and cleared the mechanical obstruction, and the line was restarted. It was all very impressive, and it is notable that although it was a mechanical fault the electrician overlooked demarcation disputes. Such production lines are very inflexible and the introduction of self-repairing robot systems is urgently required if the over-all reliability is to be improved.

Typical of the effect of the introduction of a completely automatic production facility is that quoted by Vladievskii[58]. The operating staff required for an automatic line for the manufacture of pistons was reduced from 115 down to only 6, with the output increased from 17 000 to 23 000 per worker and the cost per piston reduced from 0.96 roubles to 0.83 roubles. However, although the actual operating staff was drastically reduced, the total number of workers required was only reduced from 123 down to 106; hence the relatively small increase of output per worker compared with the great reduction of actual operating staff. The point was made that in fact physical human labour is being replaced on such a line by mental human labour and that the human facilities are thus used more effectively.

The number of letters and parcels handled in the mail is phenomenal, and various devices are being introduced which can not only reduce the number of humans required but also greatly increase the speed of handling. Automatic segregators separate letters from parcels. Automatic letter facers turn all envelopes around so that stamps are all in the same position, and then they sort out first- from second-class mail and cancel the stamps. Letters are coded, at present by humans reading the post-code or the written address, and then they are sorted according to the destination codes printed on them in phosphorescent dots. The required human intervention is reduced to a minimum. A system proposed for use in Israel even avoids the town post-code by simply asking the sender to mark a dot on a map to indicate the destination.

Robot passenger-carrying vehicles

Robot devices are put into operation in some applications in which the general public never even notice that the robot has taken over. On the new Victoria Line subway in London the trains drive themselves[8, 9]. This in itself is not new[22], because for many years the Post Office has used a driverless subway train system in London to carry the mail between sorting offices[23, 34]. However, now the reliability of the robot train is at last trusted to the extent that, although a single human staffs the Victoria Line trains, his function is merely one of supervision. There are automatic controls to start and to stop the trains, and to control the acceleration and the deceleration. If the signals are set at danger, then the train automatically stops, restarting only when the signal indicates that the track ahead is clear. The human signalmen have been replaced on this line by robot programme machines which route all of the trains throughout the week in accordance with a timetable. The central control room of this line, at Euston, is staffed by two men. However, they only intervene and take over control in an emergency. Otherwise their task is that of observing the operation of the robot system.

The passengers do not need to think about the revolutionary nature of the system which is carrying them. It has been known for many years that the robot passenger-carrying train was a possibility, and now at long last it is with us. How long will it be before it is decided that a human 'driver' on each train is unnecessary[56, 57, 73, 74]?

In Vietnam an unmanned rail vehicle has been used to proceed ahead of a passenger train, for both track inspection and protection against sabotage[60]. The robot can explode mines and signal any trouble back to the driver of the following train. It can travel at a speed of 65 km/h. The vehicle is in partial control from the following train, the amateur frequencies of 144 MHz being used for command and 27 MHz for warning signals. Other applications have also been suggested for this development, some in more peaceful fields[61]. In the writer's experience, a passenger-carrying trolley on an overhead track driven by a ropeway has been remotely controlled at about 60 MHz for oil-drilling in the ocean.

An early form of vehicular-mounted mine detector applied the brakes of a Jeep automatically if a mine was detected in the road[65]. A mobile robot device known as Joshua, which has been used in Northern Ireland, is remotely controlled by signals carried by flexible wires. It if is suspected that an abandoned automobile has been booby-trapped, Joshua trundles up to the car, smashes the window and drops in an explosive package which can destroy the car together with any deadly contents.

There are occasional proposals for the robot control of passenger-carrying highway vehicles[25, 25a, 26, 29, 36, 41-43]. However, it appears somewhat premature to discuss the method of control of a driverless but presumably passenger-carrying car at speeds of 50 km/h along a road 12 m wide. Such a vehicle moves 14 m, i.e. more than the width of the road, in each second, and in controlling such a vehicle a human being must look as far ahead as possible if disaster is to be avoided: a theoretical study should take this into account. A high speed of response is a poor assumption to make in such a case, since it not only ignores the practical possibility of construction of an actual mechanism having such a rapid controlled movement but also ignores the nature of the contact between the wheels and the road. In wet weather control can be lost, as we all know.

Copying robots

One of the most significant advances in recent years in the application of robot devices is the way in which computers are now being used to copy the actions of human operators of machines. For example, in a steel rolling mill the human operator has to control the screwing down of the steel rollers between which the steel billet passes. This control is a very skilled job, since there must be as rapid a reduction in thickness of the steel as possible before it cools. If too great a reduction is attempted, and the rolls are adjusted to be too close together, then the steel cannot pass, an irresistible force meets an immovable object and the resulting mess takes an expensively long time to clear up.

Now a computer can be adjusted to watch the actions of the human operator over several months and to record the successful actions while ignoring the unsuccessful ones. The computer can then take over from the human operator. We are left with a very real human problem: What happens to the now-redundant human? It is an urgent fact that we have to learn very quickly indeed now how to cope with this sort of problem.

Moving robot devices are in fact being used at the present time. Not only are mechanical hands being used to load and unload machine tools, but mobile robot devices, such as the Robotug[31, 32], are making their appearance in industry to transport goods from one point to another in accordance with human instructions.

The Golem waits only for money to give it life. And with the necessity there, the money will appear. We have only to think of that well-known robot, the autopilot used in aircraft, to see that this is true.

When the robots do come to help us, we shall eventually notice them no more than we now notice the wonder of the automatic washing machine or the dish-washer or the automatic record-changer on our gramophones: because it will be when they can be made in large quantities and so made economically. The more general-purpose our robots, the cheaper they will be.

We shall not notice them—except when they break down and have to be repaired. Then we shall realise how important they have become in our everyday lives.

Robot economics

The experience of the writer in developing, in producing and in selling automatic electronic control devices has led him to the conclusion that there are many people in industry who either do not understand the nature of the costs in their factory or, more likely, are not allowed to consider them fully because of the internal political considerations within the organisation which employs them.

At the level at which decisions are made on which is the best form of equipment to be employed for a particular task, it is usual for only a very short-term view to be taken. There often seems to be an undue emphasis on the initial capital cost of equipment. Other considerations, such as reliability, on the other hand, have to take a much lower order of precedence in the selection of equipment, even though these other factors are often associated with large and persisting repetitive overhead costs, such as maintenance labour and down-time of production machinery. The running cost of machinery and equipment will be an all-important factor in considering the over-all economics of operation. How often is this obvious point ignored completely when the question of immediate expenditure of money is considered, compared with the long-term cost of running and maintaining the equipment. One formal and logical method of assessment of proposed new equipment for any particular application is discussed later (p. 275).

It is, however, of interest at this stage to consider briefly the possible economic implications of the introduction into industrial operations of robot equipment. It is usually possible these days to rent expensive equipment, and so to reduce capital investment, though of course the interest rate equivalent is usually high. Even if the equipment is rented and is at the same time only used on a single-shift basis, there is for some operations an over-all reduction of operating costs compared with the cost of human labour. This is particularly true where the job is in any case rather degrading for a human to do and the rate of rejected parts produced by a human operator is consequently high.

If rented robot equipment is used on a three-shift basis, the over-all costs of operating and capital depreciation, etc., can be as low as one-third of that of a human operator, even assuming that a single robot is incapable of performing the duties of more than one man. If the capital is available for outright purchase of robot equipment on an assumed ten-year pay-off basis, then it is possible to reduce the over-all robot cost to one-fifth of that of a human operator, again assuming only a one-to-one equivalence.

No forward-looking industry can afford to overlook such figures, even if it is only to consider them carefully as applied to their own situation and then to refute them completely in their particular case. No body of shareholders can afford to allow their board to ignore such information. No forward-looking trade union can afford to overlook such figures and then not to do its best to ensure that robot methods are introduced wherever possible to reduce any drudgery of its members, while at the same time ensuring that its members share in the financial benefits and that they certainly do not suffer financially or otherwise from this inexorable next stage of progress.*

* A spokesman for the United Steel Workers of America has been quoted as saying: 'In general we don't oppose the use of these devices. We don't believe anyone can stand in the way of this kind of progress'[21].

Unfortunately, the progress can now be extremely rapid. From California we learn of the robot grape picker which replaces 40 workers. However, the mechanical device can work round the clock, so it can actually replace 100 workers or more—presumably poor migrant workers, who have to rely on the availability of this work in order to live.

An American blueberry-picking machine has hydraulically controlled fingers which shake off the berries by being vibrated with an amplitude of around 3 cm at a frequency variable from 600 to 1350 cycles per minute[75]. A British blackcurrant-picking robot, powered either by a tractor or by its own 25 kW engine, vibrates the branches 1500 times per minute with an amplitude of 5 cm in each direction and a slow rotation caused by the vibration of heavy weights. Three independent shakers in sequence direct the currants into boxes at the rear. The vehicle does require three driver–operators, but it saves £10 per tonne since it replaces some 500 human pickers. Let us hope that we can temper our progress with mercy!

Many trade unions are alive to the dangers[75]. For example, an American company planning to introduce a laser method of cutting cloth for garments, controlled by a computer, decided to retrain any workers who are displaced. The leaders of the relevant trade union were prepared to accept the machine as long as workers were not laid off or forced to take wage cuts[10, 35]. This is a good move, but it should be recognised that it is a short-term move, since there will be no need to take on new labour as the older workers retire or leave.

The robots, the general-purpose robots, are coming. Surely we have enough hard-won knowledge of economics today to ensure that the misery which the writer saw, and remembers only too well, in the 1930s cannot be repeated. And yet we must take care about predictions by economists, some of whom are too prepared to extrapolate on their curves of past history. It is less than one decade since the writer attended lectures by a very eminent economist, responsible for giving advice to important industrial leaders and for writing their speeches. This expert predicted very confidently that the population of the British Isles would be between 30 and 40 million by the end of the century, basing his prediction on extrapolation, on expected emigration from this island and on the industrial outlook for the country. Well, we have only 30 years left to prove him wrong!

The notes taken from the lectures by the eminent economist are as follows:

'There are differences of opinion on possible future population trends:

(a) The mortality and fertility rates could fall. The latter did fall in the 1930s. The worst forecast on this basis is a population of 15 millions in 2050.
(b) The replacement fertility rate is about 2.7. The population, if this is maintained, will be about 45 millions in 2050.
(c) Many economists fix a compromise figure of 30 millions in 2050.

The possible effect of widespread emigration later in this century should not be overlooked.'

In the 1969 edition of the official C.O.I. handbook *Britain* the estimate is 74 574 000 by the year 2000. (In the more recent official 'Population Projections No. 2' of 1972, the prediction is over 66 millions by the year 2011.)

I feel that we must beware of economic experts who are not sufficiently versed in the possibilities of engineering to be able to see clearly into the future. A return to the conditions of the 1930s is only just around the technological corner, unless we take care. For the problem now is one of speed. We do not now have very long before the general-purpose robot becomes available at a very low price, mass-produced like television sets or washing machines, and, because of its low cost, begins to be introduced widely into industry.

Even now in certain industries workers are given a 13 week vacation every 5 years[45]; and estimates[46] of a 7.5 hour day, 4 days per week, 39 weeks per year by the year 2000 already seem to be dated by current trends in robotics.

There is going to be much more free time. Yet such surveys as there have been seem to indicate that greater time free from work is placed last in the list of preferences expressed by workers[47, 48]. If we are wise, we will be preparing for the future age of leisure even now, so that when it becomes our turn to be declared redundant we are not affected too traumatically.

Robots and redundancy

There is a widespread, though half-hidden, fear of the future effects of the introduction of robots into industry and into commerce. Will it inevitably bring widespread redundancy and unemployment? The most likely answer at the present time is 'Yes!'

In the past the introduction of new techniques such as mechanisation and automation of particular processes has always made possible a reduction of the labour force requirement, and there is no reason to think that this trend will not continue in the future. However, in the recent past the introduction of any new techniques has been relatively gradual and has coincided with a gradual expansion of industry. It has consequently been possible to reabsorb the redundant work people.

The people themselves have to some extent helped to slow down the introduction of new techniques by both official and unofficial industrial action. This course has been more prevalent in some countries than in others, and it has been a particular feature of those countries with a long, and sometimes bitter, industrial history. It has also depended on the preparedness for militant industrial action on the part of workers, whether organised in official trades unions or not. This is particularly true where there has been a history of considerable suffering caused by widespread long-term unemployment in the past. In such a case there is an understandable anxiety that past events shall not be repeated.

An additional important point is that the process has been slowed down by the sheer capital cost of the introduction of the special-purpose, single-task, automatic machines which have had to be designed and used. Now, however, as far as can be predicted, the general-purpose robot, capable of performing not only one but a whole variety of different tasks in industry and in commerce, is about to appear. When this happens, it will no longer be true that replacement of human by machine labour necessitates an excessive capital expenditure.

Present estimates indicate that production versions of machines of the Astra type[11, 12] being investigated at Aston, to be described later, will sell at about the same price as a colour television set. Even these simple learning machines will be able to copy the control actions of a human operator controlling, for example, a lathe or a conveyor system. Future Astra machines, now being developed, will be capable of a much more complex variety of action, and they will be able to take more note of the requirements of humans and of the environment. And they will be mobile robots.

What are we to do when these general-purpose robot devices appear as low-cost capital items on the market as suggested above? Many people are likely to be unemployed or perhaps, with a more enlightened industrial policy, underemployed. In the past, this has implied 'poor'. Anthony B. Connole, of the United Auto Workers Union of the United States, has been quoted as commenting, 'Robots don't take coffee breaks, don't belong to Unions, never complain and can work long hours without getting tired or paid. It is also just as true that they don't buy anything.'[53] This profound remark should be pondered on long and taken to heart by any person in industry who is faced with the decision of whether or not to declare any employee redundant. In a Welfare State the problem tends to be obscured, but no industry can prosper unless it has an eager body of very many prosperous consumers. This must apply no matter, whether the political system is capitalist, socialist or communist.

Will there then be widespread poverty in our future robot world, even in the West? There is certainly no necessity for the robot to bring misery: with low-cost robot labour largely replacing human labour, wherever this is in any way repetitive, it will be quite possible to reduce the selling price of manufactured goods relative to the average cost of human labour. In such circumstances a reduction of the average human working week will be not only possible but essential. However, let no one think that the process will be free from industrial unrest.

If we are wise, we should be planning for this future life now, for we cannot stop it from happening. At the very least, we should be planning even now for the increased leisure time which is going to be an inevitable feature of human life in the not-too-distant future.

There is, however, one point about the robot that never seems to be stressed. A really general-purpose robot cannot be expected to learn to perform any given complex task successfully in any shorter time than is taken by a really general-purpose human child. This point might give some comfort to those who are over-worried about the coming robotics revolution. At the same time, the robot can spend twenty-four hours a day in its learning, and it can be taught by an already skilled robot.

An additional important point is that it will not be difficult with most forms of robot brain simply to erase all past memories at will. Often, as the French physiologist Claude Bernard reminded us, it is what we think we know that prevents us from learning, and the robot can be made as free from this human tendency as required.

At the present time there are very few of the existing forms of expensive robot loading device in use in industry. Even General Motors has fewer than 30, engaged on welding operations. In these circumstances the human worker has little to fear, and he is able to refer to the robot on the production floor affectionately as 'Heathcliff' or 'Iron Charlie' or 'Chezley' or 'Clyde the Claw'. When a robot in a stampings factory near to Chicago went wrong, the human workers sent flowers and 'Get Well Soon' cards to 'Sick Clyde'. Will the attitudes change when the number of robots increases to the point where human jobs are threatened in larger numbers? Or will the possibilities for increased leisure with no reduction of income be exploited and welcomed?

Robots for test and inspection

In addition to being used for domestic service and for manufacturing processes, the robot can also be used to carry out the test and inspection function in manufacture[7, 13-17, 33, 44, 59, 63, 64, 67]. Indeed, in such an application there is a great advantage, since even in very repetitive inspection the robot will not become inefficient because of tiredness. The weariness of some human inspectors in industry is notorious. Indeed, in batch production if one has some doubt about the quality of one's work, experience shows that it can be advantageous to present the work for inspection just before the afternoon tea break. (Any later and it might be put off until the next morning, when the inspector will again be fresh.) The robot inspector is not subject to such human variability.

Typical of the possibilities of automatic test and inspection is the investigation[18] carried out by H.W.C. Pilling and the writer into the possibilities of the automatic manufacture of H.R.C. fuse links. In this instance the test and inspection function is vital if there is to be a feedback of information to the manufacturing machines. Another typical example was the equipment designed by the writer for carrying out a whole sequence of tests automatically on electromagnetic contactors. This included automatic tests for noisy operation, the human tester being noticably variable in such testing unless he is given special equipment for measuring the audio noise.

There is a need for robot test equipment in automatic repetitive life testing of manufactured equipment. An early example with which the writer was concerned was equipment for the life testing of dry cells.

Another typical tester was intended for repetitive life testing of a new design of time delay relay which incorporates semiconductors. This testing device[19] was designed by B. W. Jarvis and the writer to replace an apprentice, who had carried out this form of testing previously. The time delay relay to be tested is connected to the tester, which then applies a supply voltage. At the same time, the tester starts a counting timer. After a fixed time, determined by the setting of the relay, the relay should operate. If it does, the tester leaves the supply switched on for a short time; then the supply is removed for a short time; and then the testing cycle is repeated. However, if the time delay relay being tested operates more than one-fiftieth of a second early or more than one-fiftieth of a second late, then the tester detects this fact, stops the test, records the faulty time at which the relay did operate and sounds an alarm in order to call a human to investigate the fault.

It is noteworthy that this equipment was instrumental in discovering an almost random fault in an early form of semiconductor timer, a fault which had not been spotted in the course of earlier preliminary life testing using an apprentice as the operator. Because of this achievement, the device became known as the 'Automatic Apprentice'.

Automatic testing equipment is now becoming very important in industry, particularly as the complexity of manufactured apparatus increases and at the same time the cost of the necessary skilled human labour for testing also increases. A whole sequence of tests can be carried out in a minute or so on complex electronic equipment, and the test programme can be changed simply by plugging-in a different instruction card to the automatic tester. If any fault is detected, the tester stops and indicates the nature of the fault[76].

In order to give a reliable broadcasting service, it is necessary to have some 70 separate transmitting stations, in Britain alone, which can operate semi-automatically—that is, without continuous human supervision. The Association of Broadcasting Staff was fully consulted by the B.B.C. when the automatic systems were developed, and it was realised that they relieve staff for the performance of work of greater responsibility, where they can gain more experience and enhance their prospects of promotion[68].

Robot devices are now being used to relieve nurses from routine patient-monitoring in hospitals[50, 70].

Robots for poor environments[71]

There are some applications for which robots would be ideally suited because the environment is not satisfactory for the survival of a human being. For example, there is no necessity for a robot to take in oxygen from its surroundings. Consequently, robots are suitable for use under water or in vacuous space or in an obnoxious atmosphere.

Mobile robot devices are being actively designed and used in the exploration of the moon and of the planets, and non-mobile devices such as Surveyor[49] are already in use on the moon. When it comes to the exploration of Jupiter, no man could survive there, but a robot device might do so.

There is a need for such devices much nearer home—for example, in the examination and repair of sewers. There are many environments which are too hot for human survival. The robot fire-fighter, a machine which not only detects fire but also extinguishes it, is even now under active investigation[72]. At the present time many people still suffer from ill-health caused by working in those environments in industry which are either too hot or too cold or too dirty or dusty or dangerous for human beings. On humanitarian grounds alone, robots are most urgently needed here.

In coal mining, for example, the present tendency to increased mechanisation will almost certainly lead eventually to the completely robot miner. An example of a mining machine is the automatic coal cutter, which maintains a certain thickness of coal above the roof to protect against falls of friable rock. A radioactive coal sensor using a caesium iodide source and a photomultiplier receiver pressed against the roof is used in the control, and it is possible to maintain constant the unseen thickness of the coal. It is estimated that the extra coal obtained by such a machine costing some £8000 can be worth about £130 000 per year. Success with such supervised robot miners is leading to the development of more advanced machines such as the double-ended Nicodemus[69].

The writer has found in industry that it is sometimes possible to make use of very simple electronic devices to enable a human being to be relieved from working in an environment which is so dirty as to be barely tolerable. It is no longer necessary that human feet have the skin burned from them in the heat of a brick kiln as it is being unloaded. It will soon not be necessary for any human being to dress up in shiny heat-reflecting clothing to enable him to approach close enough to tap the molten metal from a blast furnace.

Much of the work on robot hands and limbs is encouraged by the need for such devices in the handling of radioactive materials and of explosives. The term 'telechiric' has sometimes been applied to such devices. It is derived from the Greek for 'distant hand'.

In some cases it is possible to have a robot type of device remotely controlled electrically by a human operator. Unfortunately, in other cases it is difficult or impossible. For example, the control of a robot device on a distant planet would be very difficult if the time taken for the control signals to travel from earth to the planet at the speed of light is several seconds and the time taken for the information about the result of the action to return to earth is again several seconds[20, 30, 40, 51, 52].

Some forms of robot device have the advantage that they can operate completely in the dark. For example, there is no need for the tunnel in which a robot train carries mail to be illuminated at all. Even now, mobile robots are being made to inspect the inside of small diameter drain and oil pipes up to 14 km in length[54].

Robot senses

Our robots must have certain sense organs in order to be able to carry out their tasks successfully without human intervention. Some of these will simulate human senses, such as hearing, sight and touch. However, with the robot we need not be limited to the human senses. A robot can be made to detect directly radio waves or ultrasonic vibration or ultra-violet light or electrical signals, simply by the connection to its central nervous system of the appropriate sensors giving a suitable electrical output signal.

A robot anaesthetist in a hospital can directly control the amount of anaesthetic supplied in accordance with the level of the electrical rhythms of the human brain, as detected by electroencephalograph electrodes on the patient's scalp[62].

Present work on robot sensing devices is helped by the need to provide interfaces for our digital computers. Various forms of device are being tried all over the world. Pattern recognition devices, artificial eyes for reading numbers and letters, artificial ears for detecting speech, and various touch sensors are among the sense organs being investigated. It is hoped that some of this work will help to ease the task of communication between the human and his computers and his other machines. Certainly any such devices which are developed will later be of use in our robots.

There are animal sense organs which we do not yet know how to duplicate. Outstanding among these are the senses of smell and of taste. Fortunately, at the present time these senses are not very important. No doubt if they were, the required research into methods of duplication of these senses would be carried out.

It will be useful if the sense organs of a robot are arranged to have various properties which are possessed by animal and human sense organs. For example, the phenomenon of accommodation is a most useful feature. When there is a sudden change in the stimulation of an animal nerve cell, the rate of output of nervous impulses to the nervous system increases rapidly. However, if the new stimulation is maintained at an unchanged value, the nervous activity dies down. An example of this form of activity is shown in Figure 2.1 (p. 25). In effect, there is a form of quasi-differentiation in response to any change of action. Such a form of activity appears to be basic to the animal nervous system, and it is not restricted to operation in the time domain. For example, it appears that in the retina of the eye there is a form of spatial quasi-differentiation which has the effect of emphasising the edges of any image falling on the retina. It is desirable that such time- and space-differentiation action be introduced in robot nerve sensors.

A further desirable property of the animal nerve sensor is that the output is proportional to the logarithm of the magnirute of stimulation, rather than directly proportional to the magnitue. This law (known as the Weber–Fechner law) makes it possible for animal sensors to operate over a very wide range of stimulation intensities, and it is therefore a desirable feature for introduction into the system of any robot.

An interesting point is whether or not the robot should have an inherent sense of time. Experiments on sensory deprivation of human beings make it appear possible that the human in fact judges time entirely by comparison with external events and that in the absence of incoming information from external sources all time sense is lost. Even if this is the case, it will not be difficult to fit the standard robot with some form of watch or other time-keeping device, and this would be a most useful addition, since the robot could then be instructed to carry out a certain action at a particular time.

REFERENCES

1. Young, John F., *Cybernetics*, Iliffe (1969).
2. Thring, M. W., 'Automation in the Home', *Electron. & Power*, **14**, November, 440 (1968).
3. Asimov, I., 'The Perfect Machine', *Sci. J.*, **4**, October, 115 (1968).
4. Shelley, M. W., *Frankenstein*.
5. Capek, K., *R.U.R.* ('Rossum's Universal Robots').
6. Asimov, l., 'The Caves of Steel', *Galaxy* No. 13, 112.
7. Bearcroft, K., 'Automatic Testing and Automatic Test Systems for Communications Systems', *Trans. IEEE*, **COM20**, October, 1029 (1972).
8. Robbins, R. M., 'The Victoria Line and its Successors', *Electron. & Power*, **16**, June, 226 (1970).
9. Smith, V. H., 'Automation on the Victoria Line', *Contr. & Instrum.*, June, 37 (1970).
10. Anon., 'Cutting Cloth by Laser', *Time*, March 22, 19 (1971).
11. Young, John F., 'Electronic Learning Machine', British Patent Application 30657/68. Sponsored by N.R.D.C.
12. Young, John F., *Cybernetic Engineering*, Butterworths (1972).
13. Anon., 'The Automation of Testing', *IEE Conference* No. 91, September, (1972).
14. Huggins, P., 'Statistical Computers as Applied to Industrial Control', *J. Br. IRE*, **14**, July, 309 (1954).
15. Sargrove, J. A. and Huggins, P., 'Automatic Inspection—The Anatomy of Conscious Machines', *J. Instn Prod. Engrs*, **34**, September 563 (1955).
16. Sargrove, J. A. and Johnston, D. L., 'Automatic Inspection as a Key Element in Automation', *J. Br. IRE*, **17**, October, 529 (1957).
17. Sargrove, J. A., 'Automatic Inspection-Cybernetic Machines', *J. Br. IRE*, **24**, September, 241 (1962).
18. Young, John F. and Pilling, H. W. C., *G.E.C. Internal Rep.*, SWD 18 (1957).
19. Young, John F. and Jarvis, B. W., *G.E.C. Internal Rep.*, SWD 34 and SWD 44 (1958).
20. Leslie, J. M. and Thompson, J. A., 'Human Frequency Response as a Function of Visual Feedback Delay', *Hum. Factors*, **10**, 67 (1968).
21. Rosenblatt, A., 'Robots are Ready to Grapple with Dirty Jobs in Factories', *Electronics*, **40**, March 20, 165 (1967).
22. Dell, R. and Manser, A. W., 'Automatic Driving of Passenger Trains on London Transport', *Proc. Instn Mech. Engrs*, **179**, pt 3A, September (Autom. Rlys Special), 24 (1964).
23. Mew, G. M., 'The Post Office Railway', *Proc. Instn Mech. Engrs*, **179**, pt 3A, September (Autom. Rlys Special), 39 (1964).
24. Kohl, R., 'Adaptive Control: Toward the Thinking Machine', *Mach. Des.*, **41**, May 1, 156 (1969).
25. Bierman, G. R. and Hain, J. L., 'The Automatic Highway', *Mech. Eng.*, **90**, July, 18 (1968).
25a Flory, L. E., 'Electronic Control of Highway Vehicles', *JIEE*, **7**, May, 271 (1971).
26. Gaines, B. R. and Andreae, J. H., 'A Learning Machine in the Context of the General Control Problem', *I.F.A.C.*, June 21, paper 14B (1966).
27. Brewer, R. C., 'The Numerical Control of Machine Tools', *Engrs' Dig.*, *Lond.*, *Supp.* (September, 1959).
28. Brewer, R. C., 'Recent Developments in the Numerical Control of Machine Tools', *Engrs' Dig., Lond.*, **22**, August, 91 (1961).

29. Bender, J. G. and Fenton, R. E., A 'Study of Automatic Car Following,' *Trans. IEEE*, **VT18**, November, 134 (1969).
30. Ferrell, W. R., 'Remote Manipulation with Transmission Delay', *Trans. IEEE*, **HFE6**, September, 24 (1965).
31. Helps, F. G., 'Driverless Tractor for Materials Handling', *J. Br. IRE*, **25**, March, 273 (1963).
32. Anon., 'Driverless Tractor Trains', *Rly Gaz.*, **123**, 36 (1967).
33. Sargrove, J. A., 'Automatic Inspection', in: Booth, A. D. (ed.), *Progress in Automation*, Vol. 1, 209, Butterworths (1960).
34. Rossion, M., 'Transportation of Express Mail to the Sorting Post Office at Brussels South Station', *ACEC Rev.*, No. 3, 24 (1959).
35. Anon., 'Computers in the Cutting Room', *Control Eng.*, **16**, October, 115 (1969).
36. Anon., Radar Controlled Car: A Step Toward Automated Highways', *Ind. Wk*, 172, February 14, 65 (1972).
37. Thring, M. W., 'A Robot about the House', *New Scient.*, **22**, April 2, 19 (1964).
38. Thring, M. W., 'Man's Need for Machine Slaves', *Physics Bull.*, **21**, 449 (1970).
39. Anon., 'He Envisages a Creative Society in Which Robots Do Routine Work', *Product Eng.*, **41**, September 28, 24 (1970).
40. Ferrell, W. R. and Sheridan, T. B., 'Supervisory Control of Remote Manipulation', *IEEE Spectrum*, **4**, October, 81 (1967).
41. Rosen, D. A., *et al.*, 'An Electronic Route-Guidance System for Highway Vehicles', *Trans. IEEE*, **VT19**, February, 143 (1970).
42. Fenton, R. E., 'Automatic Vehicle Guidance and Control, A State-of-the-Art Survey', *Trans. IEEE*, **VT19**, February, 153 (1970).
43. Ishii, T., *et al.*, 'Computer-Controlled Minicar System In Expo 70; An Experiment in a New Personal Urban Transport System', *Trans. IEEE*, **VT21**, August, 77 (1972).
44. McAleer, H. T., 'A Look at Automatic Testing', *Spectrum*, **8**, May, 59 (1971).
45. Klausher, W. J., 'An Experiment in Leisure', *Sci. J.*, **4**, June, 81 (1968).
46. Kahn, H. and Wiener, A. J., *The Year 2000*, Macmillan (1967).
47. Faunce, W. A., 'Automation and Leisure', in: Jacobson, H. and Roucek, J. S. (eds.), *Automation and Society*, Philosophical Library, N.Y. (1959).
48. Neuloh, O., 'Automation and Leisure', *Sci. J.*, **4**, January, 79 (1968).
49. Anon., 'Surveyor 1: Preliminary Results', *Science, N.Y.*, **152**, 1737 (1966).
50. Anon., 'Enter the Robot Nurse', *Des. Electron.*, **8**, June, 57 (1971).
51. Sheridan, T. B. and Ferrell, W. R., 'Remote Manipulative Control with Transmission Delay', *Trans. IRE*, **HFE4**, September, 25 (1963).
52. Smith, W. M., *et al.*, 'Delayed Visual Feedback and Behaviour', *Science, N.Y.*, **132**, 1013 (1960).
53. Spector, L. F., 'The Robots are Coming—or are They?', *Mach. Des.*, **42**, July 9, 38 (1970).
54. Anon., 'Pipe Inspecting Robot', *Mach. Des.*, **42**, March 5, 40 (1970).
55. Thring, M. W., 'The Robot Age', *Engineering, Lond.*, **209**, February 6, 128 (1970).
56. Guieyesse, L., 'Experience with Automatic Driving of Trains on the Paris Metro', *IEE Conf. Perf. Electrified Elys, October, 1968*.
57. Tappert, H., *et al.*, 'Automating the Hamburg Underground', *IEE Conf. Perf. Electrified Rlys, October, 1968*.
58. Vladzievskii, A. P., 'First Automatic Plant and its Influence on Automation', *Machs Tool.*, **41**, 17, No. 4 (1970).

59. Sargrove, J. A. and Johnston, D. L., 'Automatic Inspection and Control', in: Grabbe, E., *et al.* (eds.), *Handbook of Automation, Computation and Control*, Vol. 3, Ch. 4, 40 (Wiley, 1961).
60. Morrison, T. R. and Jones, T. L., 'A Radio Controlled Rail Vehicle for Track Inspection and Protection', *Proc. IEEE Industry and General Applications Meet., Michigan, 1969*, 215.
61. Morrison, T. R. and Jones, T. J., 'Remote Controlled Railway Surveillance Vehicles', *Proc. IEEE Industry and General Applications Conf., Chicago*, 1970, 339.
62. Bellville, J. W. and Attura, G. M., 'How Electronics Controls Depth of Anesthesia', *Electronics*, **32**, January 30, 43 (1959).
63. Silberstein, J. M., 'Automatic Circuit Checker for Television Receivers', *Electron. Eng.*, **23**, June, 202 (1951).
64. Fichtenbaum, M., 'Computer Controlled Testing Can Be Fast and Reliable and Economical without Extensive Operator Training', *Electronics*, **43**, January 19, 82 (1970).
65. Doll, H. G., *et al.*, 'Vehicular Mounted Mine Detector', *Electronics*, **19**, January, 105 (1946).
66. Avis, J. M., 'Harvester Fingers Shake Fruit', *Hydrauls Pneum.*, **25**, June, 86 (1972).
67. Williams, R. C. G., *et al.*, 'The Design of a Universal Automatic Circuit Tester, and its Application to Mass-Production Testing', *JIEE*, **94**, pt 3, January, 20 (1947).
68. Wynn, R. T. B. and Peachey, F. A., 'The Remote and Automatic Control of Semi-Attended Broadcasting Transmitters', *Proc. IEE*, **104B**, November, 529 (1957).
69. Hartley, D., 'Automatic Steering of Mining Machines', *IEE Colloquium on Measurement and Control in Coal Mining*, May 23, 1972.
70. Harrois-Monin, F., 'Automatic Hospitals', *Science Vie*, *121*, May, 118 (1972).
71. Heginbotham, W. B., 'Reasons for Robots', *Proc. 1st Conf. Industrial Robot Technology, Nottingham, March, 1973*, 3.
72. Whitehouse, R. B., Automatic (Fire) Detection Equipment, *Electl. Rev.*, **192**, February 16, 248 (1973).
73. Friedlander, G. D., 'Bigger Bugs in Bart', *Spectrum*, **10**, March, 32 (1973).
74. Friedlander, G. D., 'A Prescription for Bart, *Spectrum*, **10**, April, 40 (1973).
75. Cooley, M. J., 'Industrial Robots: A Trade Union View of the Social and Industrial Implications', *Proc. 1st Conf. Industrial Robot Technology, Nottingham. March 1973*, 223.
76. *Conf. Automation of Testing, IEE, Keele, September, 1972.*.

2

The senses of the robot

Human sensors

The sense organs[24] possessed by the human being are of interest to the robotic engineer since in some cases it will be necessary for him to attempt to duplicate human senses as closely as possible in the robot. For example, this is necessary where a robot is to take over a human task directly.

The touch sense of the human produces four distinct sensations, and it appears that human skin contains separate receptors for these four. The separate sensations are cold, heat, pain and pressure. The sensitivities of different parts of the skin to those sensations vary; thus the area density of the various types of receptors varies over the body.

Touch is likely to be a most important form of sense organ for the robot. However, cold and heat sensing is not so important to the survival of the robot as it is to an animal, though it might be important if, for example, a particular task is to maintain human living conditions within a tolerable range.

Pain might be thought at first to be a quite unnecessary sense for a robot. However, to the extent that such a sense can give a robot warning of excessive loading or of other danger, an equivalent to pain sensing could be valuable. This is particularly so if by such sensing it can protect a human from pain.

The indispensable sense of touch to a robot will be pressure sensing. There are various methods which can be used, and standardisation is unlikely since the methods vary in their applicability.

Response of receptor nerve cells

The transient response of receptor nerve cells in the animal has been quite well studied, and it is clear that a form of phase-advanced response is given. If the stimulus is applied and then maintained at a constant value, the impulse rate of the nerve cell first jumps to a high value and then decays almost exponentially to a lower value[2]. The initial level of the frequency–time curve depends on the size of the stimulus, as does the final steady value. This form of transient response is shown in *Figure 2.1*.

There is some doubt about the form of the response when expressed as a frequency response. This is because of the non-linearity and the difficulty of assessing the output when expressed as a definite number of impulses per second, since the output impulses are not at all regular. However, a typical response[1] has been given as

$$\frac{27(1+s30)}{(1+s5)(1+s0.35)}$$

This response gives a phase lead of about 20° over a ten-to-one frequency range centred on about 0.1 Hz.

It is certainly worth while including some provision for phase-advance in robot receptor circuits, though the form of this will depend on the over-all design of the control system when considered as a closed loop including the external environment in the feedback path.

Human neuron numbers

Since the robot is intended to duplicate at least some features of human nervous activity, it is useful to tabulate the numbers of neurons used for various purposes in the human[3]. (See Table 2.1.)

Table 2.1

	Sensory inputs			Central nervous system	Conscious output
	Receptors	Nerve fibres	Channel capacity		
Vision	2×10^8	2×10^6	5×10^7		
Hearing	3×10^4	2×10^4	4×10^4		
Touch	5×10^5	1×10^4			
Pain	3×10^6				
Heat	1×10^4	1×10^6		1×10^{10}	50 bits/s
Cold	1×10^5				
Smell	1×10^7	2×10^3			
Taste	1×10^7	2×10^3			
Totals	3×10^8	3×10^6			

24 The Senses of the Robot

It should be noted that values such as these cannot be quoted with any great accuracy, and they are reproduced here for convenience in comparison rather than for concrete purposes of engineering design. There is no attempt at accurate summation. Note that the product of the estimated maximum conscious output of 50 bits/s, when multiplied by an average life expectancy of say 70 years, gives a figure of the order of 10^{10}, i.e. the same order as the number of cells in the brain. It is also of interest to note that on average there are about 100 receptors to each nerve fibre and 3000 neurons in the central nervous system to each nerve fibre. Keidel[36] suggests that human output is 10^7 bits/s.

It has been suggested by McCulloch[53] that in robots we cannot afford to carry out any computations, no matter how simple, in a hundred parallel paths and demand coincidence, as is done in the animal nervous system. In consequence, it is unlikely that any robot computer will operate correctly under conditions as varied as those which the human body has to withstand.

By demanding coincidence between the activity of nerve fibres we vastly increase the probability that what is present in the output really does correspond to something in the input and not merely to noise or to interference. The body pays for certainty with information. The eye relays to the brain only one-hundredth part of the information which it receives. However, the chance that the information which it does relay is merely due to chance is fantastically small—1 in 2^{100}.

This is the real advantage of the one-hundred-to-one corruption in fineness of detail in passing from nerve to brawn.

It is of interest to compare the visual system of the human with the auditory system as follows[4, 5]:

	Visual	Auditory
Receptors per fibre	130 : 1	1 : 1
Neurons in primary cortex	538×10^6	100×10^6
Fibres per cortical cell	1 : 538	1 : 3300

Another point of interest is the number of different sensory signal levels which can be transmitted to the central nervous system from the various peripheral sensors and so distinguished by the human. The number varies in general between 3 and 9 different levels, though higher values are possible[28].

It is an interesting point that various forms of prosthetic sensor device have been adopted for use by human beings, and some of these methods might be of interest for use with robots. As examples, the use of the Braille alphabet for communication with the blind and the use of vibrotactile devices for communication through the skin touch sensors[34] can be quoted.

Reflex action in the robot

The function of reflex activity in the human and the animal appears to be directly that of survival. It is desirable that the robot should similarly be fitted with certain reflex actions designed to ensure the

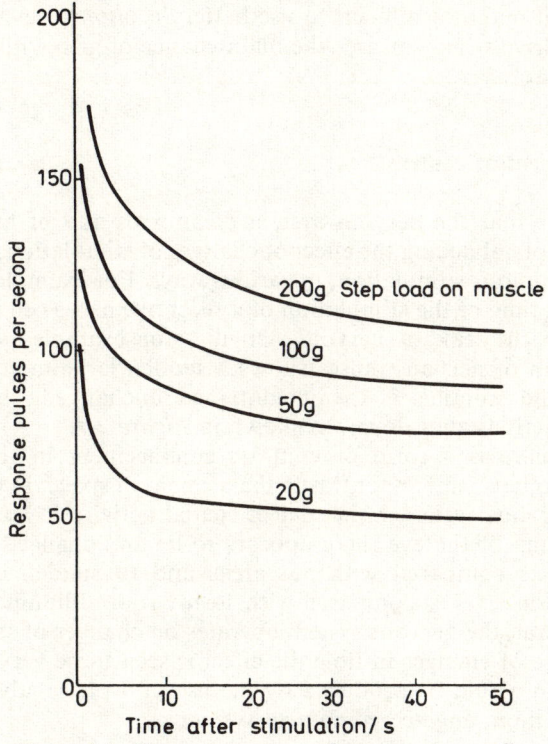

Figure 2.1 Transient response of receptor nerve cells

survival of the robot. As an example, heat sensors in the robot hand can operate rapidly and directly into the limb-motor system so that a limb-withdrawal reflex can be initiated to give rapid protection against damage. There is no need for such a reflex action to take a complete path through the robot nervous system, since the most rapid possible response is required for protection. As explained elsewhere, such robot reflex action can be made capable of association in the central nervous system of the robot with other, non-reflex, inputs to the nervous system[59].

There is, however, an important restriction of possible reflex activity in the robot which is not present in the animal system. If the tentative laws of robotics are to be followed, then it must be ensured that the reflex actions of a robot cannot harm a human being. It is better that the robot be damaged than that a human be harmed by the reflex action of the robot. This will not be easy to ensure in all cases.

It should be noted that the provision of means for rapid retractive movement makes it difficult to use battery economy devices such as a worm drive which ensures the maintenance of grip with no energy expenditure[6].

Enhancement of contrast

It appears that the nervous system of animals and of humans has the effect of enhancing the effect of changes of stimulation, compared with continuous stimulation, in various ways. For example, a sudden change in time of the stimulation of a receptive nerve cell is followed by a sudden peak of nervous activity—for example, the rate of production of nervous pulses shows a sudden large increase. However, if the stimulus is then maintained unchanged, the peak of nervous activity dies down, as shown in *Figure 2.1*.

Thus there is a form of contrast enhancement in time. There appears to be a similar contrast enhancement in space, in those parts of the nervous system which detect spatial activity[54]. For example, in the retina of the eye there appears to be an enhanced sensitivity to edges as compared with flat areas and to sudden changes of illumination level as compared with steady room illumination.

In general, the nervous system operates on changes of stimulation. In the case of changes in time the effect is seen to be very similar to that which would be produced by the use of a phase-advance form of circuit in an engineered control system.

Gas and moisture detection

It is not at present known how to reproduce the human sense of smell electronically. However, it has proved possible to detect certain gases[7-9]. For example, if a combustible gas and air mixture comes into contact with certain catalysts such as platinum or palladium, then heat is generated and the electrical resistance of the catalyst changes. The change of resistance can be detected electrically. Such detectors are of course important for fire and explosion prevention. Oxygen can be detected by making use of its paramagnetic property. Some gases can be detected by making use of their thermal conductivity in a device known as a katharometer. Infra-red absorption can also be used to determine the presence of various gases. Water vapour can be detected by various forms of conductivity hygrometer. None of these methods is very suitable for use in the general-purpose robot, and it is hoped that future research will lead to improved forms of smell detector.

The human sense of taste is another sense which cannot yet be duplicated[35]. The nearest approach is afforded by the pH type of detector, but there is as yet no method really suitable for general use by the robot. The tastes which can be detected by the human can be categorised into the four basic tastes: sweet, bitter, sour, salty, which are present in varying relative amounts. Taste can therefore be regarded as a four-dimensional measurement.

With both taste and smell the information gained by the sensing is greatest when the relative probability of the taste or smell in question is low. Only the unfamiliar is noticed.

Wright has suggested[32, 33] that odours are detected by the animal nervous system on the basis of molecular vibrations in the far infra-red region.

Gases can produce colour changes in various chemical substances, and this fact is of course used in chemical analysis. It is now possible to obtain glass tubes containing various reagents through which a gas can be drawn by a simple piston and cylinder, operated by hand. Such a technique would be applicable to a robot, though here it would be preferred that the reaction be reversible and self-recovering so that it would not be necessary to change the detector tubes each time that detection has taken place. However, it should be noted that it seems that the life of the animal smell detection nerve endings is very short and replacement is very frequent.

A Japanese gas-sensing device is the Taguchi Gas Sensor[39], which is composed of oxidised metals such as tin oxide, zinc oxide and ferric sesquioxide. This device undergoes a very large, though reversible, decrease of electrical resistance when it encounters deoxidising gases such as hydrogen, carbon monoxide, methane, propane, alcohol, volatile oil and acetylene.

An oxygen analyser which can, for example, be used to determine the concentration of oxygen in flue gases and so be used in the control of combustion efficiency, has been produced using solid state technology[40]. A stabilised zirconium cell is operated at a constant temperature of 850°C, and it generates a voltage which varies logarithmically with the difference between the oxygen partial pressure and that of a reference source. Accuracies of $\pm 0.1\%$ with a response time of 0.2 s in temperatures from 10 to 760°C are claimed, with output signals of 4–20 mA or 1–5 V.

Various elements have been used for humidity sensing, including lithium chloride (the 'Dunmore' sensor), carbon, polyelectrolyte resistance elements, ceramic elements, capacitive devices[48] and aluminium oxide elements[41]. All are more or less unstable because of ionic contamination, dissolution in water, polarisation, or chemical or mechanical degradation. Thoma[42] has used a ribbon-like strip formed from a five-layer film of cellulose acetate butyrate, which is claimed to give a high sensitivity while being chemically and mechanically stable. The chemical in finely divided deoxidised carbon is used as a core, and the element is treated with an aqueous solution of sodium hydroxide to increase the sensitivity. Resistances of about 2500 Ω are obtained and operation is possible with relative humidities of 10–90%, with a time constant of more than 100 s. The resistance is little affected by changes in voltage or temperature. It is likely that developments such as this will be very useful in robotic applications.

Temperature regulation

The mechanism of all animal control systems is affected by temperature. The internal temperature of a cold-blooded (or poikilothermic) creature follows the ambient temperature variations, and because of this the level of bodily activity is limited to some extent by the ambient conditions.

On the other hand, warm-blooded creatures have automatic controls of heat loss and of heat generation which maintain the internal temperature of the animal fairly constant. The minimum permissible body temperature is about 37°C, below which enzyme activity is greatly reduced. On the other hand, the cells of the central nervous system are irreversibly damaged by subjection to temperatures above about 41°C. Thus a fairly close control of bodily temperature is required.

Fortunately, in the robot the extensive means which have had to appear for the purposes of thermoregulation during the course of evolution of the animal system are not at all necessary. Some temperature regulation is required, but the specification is not too severe. As an example, semiconductors are available which will operate at ambient temperatures in the range −50°C to +150°C. Since electrical and electronic control circuits and devices all lose energy in the form of heat, a form of cooling is often necessary. This is often the case if the equipment is compact—for example, integrated electronic circuits. However, since a close control of temperature is not necessary, elaborate temperature control systems are rarely required, and simple cooling systems using convection currents in the surrounding air are sufficient[10, 11].

It is worth noting that the heat regulation of the animal body is facilitated by the fact that there is a water content of some 70%. Because of this a given amount of heat outside the body produces far less effect on the body temperature than it would otherwise. Water has one of the highest latent heats of freezing, and so it regulates the temperature of the earth, since heat reduces the amount of surface ice while cold increases it. Similar methods have been used with other materials to maintain constant the temperature of electronic components such as quartz crystals. The latent heat of vaporisation of water is also one of the highest known, and this fact has been used in 'boiling water' cooling systems. The freezing point of water is some 100°C higher than that of the more common gases. All of these points make the water content of the animal body a very useful feature. Unfortunately, a robot with a large body-water content would be rather heavy, and this might be a disadvantage.

In some applications it will be necessary for the robot to be capable of determining temperature levels, either in its own structure or in its surroundings. In such cases any of the well-known methods of electrical temperature determination can be used[12]. For example, resistance thermometers can be incorporated if accurate measurement is required, though the electrical output obtainable is somewhat low.

Semiconductor devices are extremely sensitive to temperature. For example, thermistors made of the oxides of various metals have been quite widely used[13-15, 43]. One difficulty with some thermistors has been the slow response to changes of temperature. This has been minimised by the use of physically small thermistors, and more recently fast-response devices made of silicon have been produced.

One most ingenious method of making a thermopile consisting of a number of thermocouple junctions in series should be mentioned. This is the use of constantan wire, plated in stripes with copper. Since the copper is a better conductor of electricity, most of the current flows through the plating in the copper-covered sections, though it all has to flow through the constantan elsewhere. By this means, it is possible to obtain the effect of a very great number of junctions in series, and the arrangement is very useful for determining the flow of heat.

Temperature control using thermistors

Where a robot must be equipped for accurate control of temperatures, it is possible to incorporate devices such as resistance thermometers, together with the necessary amplifiers. However, in many applications an extremely accurate control is unnecessary and it is desirable to simplify the required equipment as much as possible. In such applications the thermistor is an attractive device. The thermistor is a semiconductor resistor which has a high negative value of temperature coefficient of resistance, a typical value being -4% per deg C at 20°C.

The resistance of a thermistor varies exponentially with temperature; thus it can be expressed by

$$R = A \exp(B/T)$$

where R is the termistor resistance; A is a constant; T is the thermistor temperature in K; and B is a constant, typically 2500–3000 K. By differentiation it follows that

$$\text{temperature coefficient of resistance} = -\frac{B}{T^2}$$

Thus the temperature coefficient decreases with an increase of temperature.

It is often desirable that the variation of resistance with temperature should be linear, at least over a limited temperature range. This can be achieved by the addition of a fixed value of shunting resistor, and a limited region of linear conductance variation with temperature can be achieved by the addition of a fixed series resistor.

The required shunt resistance R_p can be found from

$$R_p = R\left(\frac{B - 2T_i}{B + 2T_i}\right)$$

where T_i is the required median operating temperature and the subscript 'p' stands for 'parallel'. At this temperature T_i there is an inflection in the curve of total resistance against temperature. The rate of change of resistance with temperature is a maximum at this point. It is equal to

$$\frac{dR_p}{dT} = -R\frac{(B - 2T_i)^2}{4BT_i^2}$$

In a similar way the required value of series resistor to give a limited range of linear variation of conductance is obtained from the formula for required series conductance G_s

$$G_s = G\left(\frac{B + 2T_i}{B - 2T_i}\right)$$

and at the inflection point

$$\frac{dG_s}{dT} = G\frac{(B + 2T_i)^2}{4BT_i^2}$$

where G is the conductance of the thermistor at temperature T_i.

In this way linearity is obtainable over a temperature range greater than some $\pm 20°C$. A similar linear range of current variation is obtainable from thermistor bridge networks, and various rules of thumb have been devised for the design of these circuits.

As an alternative to a thermistor, a semiconductor diode can be used as a temperature sensing element. An even greater sensitivity to temperature is obtained by use of a transistor with an open-circuit base[26, 31].

Thermistor linearity, of either voltage or resistance against temperature, can be obtained over a wide range by shunting one thermistor by a series thermistor–resistor combination. This has been done in the Thermilinear components produced by the Yellow Springs Company. Linearity within a fraction of a degree is obtained over a range as wide as $-30°C$ to $+100°C$.

Force feedback sensors

In many cases it is desirable that there should be a feedback in the controlling servo system of a signal indicating the force exerted by the mechanical output. This is particularly the case where, for example, robot fingers have to grip various objects, some of which are fragile.

In work on prosthetic devices the same problem has been encountered, and here, at the present time, it seems that the best device to provide the feedback signal is a strain gauge, preferably using silicon crystals. These have a negligible hysteresis effect.

In a typical prosthetic application a signal of some 2 mV/N is required, and this is obtained by the use of a Wheatstone bridge incorporating two p-type silicon strain gauges mounted on opposite sides of a mild steel, constant-moment, beam. This arrangement takes a constant current from the supply and it gives a linear relationship between the applied strain and the output voltage. It is necessary in the circuitry to make provision for interchangeability of strain gauges in case of failure, and this is quite easy to do. In addition, it is usual to offset the control slightly so that there is a slight grip, of the order of 2 N, in the resting position. A by-pass arrangement can be fitted to protect the strain gauge beam from over-strain.

When movement into a gripped position takes place at a high velocity, it is necessary to take account of the kinetic energy of the movement and to apply reverse torque in order to shorten the braking time of the movement.

There is some possibility of using Zener diodes as hydrostatic pressure gauges[44], and this could lead to small-sized devices for robotic use.

In some early prosthetic devices mechanical contact between granules of some conducting substance such as carbon has been used for force feedback: for example, in the case of a prosthetic hand by enclosing the granules within the outer rubber skin. Such a method is inexpensive and since it is based on the principle of the well-known carbon granule telephone mouthpiece microphone, it is well-tried. Also, it can give a high value of output voltage, typically 4–10 V. Unfortunately, such a method is unlikely to be sufficiently reliable for use in the general-purpose robot.

In one form of artificial hand a number of separate sponge-carbon pads have been used, but the outer skin 'glove' of the hand has also been filled with sponge rubber, so that the pressure at all points is transmitted to the pads. The transducer pads give a linear region of response for a certain weight range but a limited upper value of response.

It is useful to have weight sensors in the wrist of a robot hand, so that the hand acts in effect as a form of weighing machine. The information from this is of use for arm movement as well as for emergency use. It is useful to provide additional sensors which are not normally pressed but which can be pressed by a human in an emergency in order to cause the pressure of the grip to be released.

There is a need for new types of force sensor for use in robot and prosthetic applications. These should have a high value of output voltage, should be rugged and insensitive to temperature variations and should have low weight. A possible approach meeting some of these requirements is the use of a high-sensitivity piezo-crystal element to which the force is transmitted by a liquid contained in the same skin as the sensor.

Whatever form a force feedback sensor takes, it is important to follow it by adequate filtering to remove signals produced by environmental factors such as excessive noise and vibration.

In some prosthetic devices it has been found advantageous to protect against the damage which can be caused in the control system in the event of accidental breakage of the strain gauges. Both electrical and mechanical methods of protection have been used, sometimes in combination. A further safeguard has sometimes been provided by an automatic switch-off of the drive motors if there is any attempt to lift an excessive load.

In the human the touch sensors are very sensitive and because they are numerous it is possible to use them to discriminate shapes. Now such a facility would be very desirable in a robot, but at present it would be very difficult to provide because of the excessive size of the available sensors. There is a need for touch sensors as small as possible and, if possible, in an integrated circuit form so that they can be used in circuitry similar to that which is needed for the retina of the robot eye. Fortunately, such a very fine touch discrimination is only likely to be needed at one point—for example, at the tip of a single robot finger. As with the retina, edge detection will be a useful feature.

Position feedback

Provision of the accurate and reliable measurement of position in servo systems is never easy, especially when, as in a robot, the equipment is required to have a minimum size and weight. The basic need for accuracy is greatly reduced if closed loop feedback control is used. Such feedback in some form is a usual feature of animal systems.

The position sensors used in man-made control systems have taken many different forms. For accurate position measurement—for example, as required in automatic mechanical measuring equipment or in automatic machine tools—a digital form of measurement is commonly adopted. Many different methods have been used for this, and these can be loosely divided into two categories.

The first type is an absolute position measuring device, which produces digital signals directly indicating the absolute position with respect to a fixed reference datum. The second type incorporates some form of counting device which summates small uniform increments in order to obtain an absolute position. The first type is subject to gross errors in the event of failure of one of the more significant figures of the measurement. The second type is subject to a permanent loss of information in the event of failure to count even a few increments. This can be obviated only by frequent re-zeroing of the measuring device, since absolute reliability can never be guaranteed.

Such accurate position measuring devices are, in general, large and costly, and on both counts they are therefore unsuitable for use in the general-purpose robot. Fortunately, if the robot is provided with visual feedback from some form of 'eye' device, the need for absolute accuracy is greatly reduced and replaced by the need for accurate position comparison. The general-purpose robot should be capable of making use of this human method of non-absolute position measurement or determination by comparison.

It should be noted that although the human and the animal make use of visual feedback, there is in addition some form of internal feedback, since otherwise the accuracy of positioning obtained with the eyes closed or in cases of blindness would be much reduced. It has been suggested that neural feedback possibly defines a particular relationship between muscle tension and muscle length, rather than forms part of a position control servo loop.

A form of precise indication of position originally introduced by the writer uses the rotation of a steel vane to produce pulses magnetically in a coil, the time position of the pulse being compared with that of another, master pulse[16]. Once the temperature drift has been eliminated by correct mounting, this method can give a great accuracy. In order to avoid the necessity for rotating parts, the writer has adopted a linearly travelling field as produced by a polyphase winding in this form of position indication.

Magnetic pulse generation

In robot applications it is important to keep the size and the weight of any mechanical measuring arrangements as small as possible. A method of measurement which has been adopted by the writer uses the pulses generated in a coil surrounding an open magnetic circuit when a piece of iron is moved past. Since the flux linking the coil must vary owing to the varying reluctance of the magnetic circuit, an e.m.f. is induced.

The instantaneous e.m.f. e induced in a coil of N turns surrounding a magentic circuit carrying a varying flux F is

$$e = -N\frac{dF}{dt}$$

If the magnetic circuit has a constant magnetomotive force M and a variable reluctance R, then

$$e = -N\frac{d(M/R)}{dt} = \frac{NM}{R^2}\frac{dR}{dt}$$

If the magnetic circuit has one fixed part and one variable part, then

$$R = K + l/Au$$

where K is the reluctance of the fixed part; l is the length of the variable part; A is the cross-sectional area of the variable part; and u is the permeability of the variable part.

Hence,

$$e = \frac{NM}{(K+l/Au)^2} \times \frac{d(l/Au)}{dt}$$

so that

$$e = \frac{NM}{(K+l/Au)^2} \left(\frac{Au(\mathrm{d}l/\mathrm{d}t) - l(\mathrm{d}Au/\mathrm{d}t)}{A^2 u^2} \right)$$

From this, if the length l is small compared with KAu, we obtain

$$e = \frac{NM}{K^2 Au} \left(\frac{\mathrm{d}l}{\mathrm{d}t} - \frac{l}{Au} \times \frac{\mathrm{d}Au}{\mathrm{d}t} \right)$$

In this ther term $\mathrm{d}l/\mathrm{d}^8$ is the rate of increase or decrease of the length of the variable parts of the magnetic circuit. Consequently, in order to maximise the induced voltage e, the magnetic circuit should preferably be designed so that, as the length decreases, the area A increases and vice versa.

This simply implies that the rate of change of reluctance should be as great as possible. As might be expected, to obtain a high value of induced voltage we require a large number of turns and a high value of magnetomotive force, a large rate of change of reluctance and a low value of reluctance in the fixed part of the magnetic circuit. In addition, the value of the product Au should be small, though this is difficult to achieve since the rate of change of reluctance must be high.

The significance of the method is that it can be used either for position indication or for speed indication and that an excellent independence of environmental conditions is obtainable with careful design. The electronic circuitry required can be very small and light if integrated circuits are used. The method was used on very early work (1953) for electronic weighing, but could not be published until much later[16]. It has also been used, for example, for tachometry[23] and commercial units are now produced.

Resistive strain gauges

In robot applications it is important to keep the size and weight of any position measuring apparatus down to a minimum. Consequently, although in some cases it is necessary to use more complex devices such as resistive or inductive potentiometers, the simple resistive strain gauge is an attractive solution where the amount of movement is limited.

Where resistive strain gauges [17-20, 58] are to be used for fairly precise position measurement, it is very desirable that a double-bridge form of feedback circuit should be adopted. This helps to overcome problems caused by, for example, temperature variations and supply line resistance variations. It is also essential to provide a path having a good heat conductivity in order to limit the temperature rise caused by the power dissipated in the resistive strain gauge element. Resistive strain gauges must be very well protected against, for example, the ingress of moisture, since the resulting change of insulation resistance can have a surprisingly large effect.

The normal type of resistive strain gauge is basically suitable for only small extensions, and it is necessary to incorporate levers where a large movement must be monitored. One possibility for incorporation in robots is the use of conductive rubber or plastic strain gauges. Although these are little used, they have the advantage that they are suitable for the measurement and control of very large extensions.

In some applications of resistive strain gauges it is possible to arrange for the base material on which the gauge is to be mounted and the resistance material of which the gauge is made to be mutually self-compensating for temperature effects[21]. In other words, the mechanical temperature coefficient of expansion of the base material is used to compensate for the electrical temperature coefficient of resistance of the material of the gauge. Gauges are available having suitable resistance temperature coefficients to compensate for the thermal expansion of most common metals.

One problem with the resistive strain gauge is the fatigue limit of the material used. Since the gauge is subject to continual flexing in use, this factor places a limit on the life obtainable. Consequently, if the gauge is glued to the body of the robot it can cause a maintenance problem. In order to maximise the life, the variation of strain should be limited in each cycle of movement.

For some applications the piezo type of element is an attractive form to use[22, 29, 30]. Since these elements are available in small cartridge package form for use in gramophone pick-ups, they are easy to use and the cost is quite low, while the life is long.

There are many other methods by which strain can be determined —for example, capacitive or inductive methods. However, in general, at the present time these are not very suitable for use in mobile robot applications, because, for example, of the requirement for a special power supply or because of susceptibility to environmental influence.

Recently introduced strain gauges have used a resistive element made from a semiconductor material such as silicon. While these are capable of providing an output voltage some 100 times greater than that from more conventional strain gauges, they can introduce other problems at the present time. For example, there can be a large change of gauge resistance with temperature, and in some cases it has been necessary to add a thermistor for temperature compensation.

The Senses of the Robot

In addition, since silicon is not very ductile, there can be problems of breakage caused by sharp bends or pressures.

The gauge factor of a strain gauge is sometimes quoted. The gauge factor is defined as

$$\text{gauge factor} = \frac{\text{fractional change of resistance}}{\text{fractional change of length}}$$

Now if the resistance is R, the resistivity of the material is ρ, and the length, cross-sectional area and total volume of the material of the gauge are L, A and V, respectively, then the resistance is given by

$$R = \frac{\rho L}{A} \, \Omega$$

$$= \frac{\rho L^2}{V} \, \Omega$$

Hence, differentiating with respect to L,

$$\frac{dR}{dL} = 2\frac{\rho L}{V}$$

$$= \frac{2R}{L}$$

Hence,

$$\text{gauge factor} = \frac{dR/R}{dL/L} = 2$$

However, in practice the volume is not truly constant, and a typical practical value of gauge factor is 2.2.

The actual gauge factor can be regarded as being made up from two terms:

$$\text{gauge factor} = (1+2\nu) + ((d\rho/\rho)/(dL/L))$$

where ν is Poisson's ratio. The first term here is a dimensional effect, ν being in the region of 0.3 for metals. In semiconductors the second term, which is sometimes called the piezo-resistive effect, is all-important[57].

Piezo-electric strain detectors have been used as touch sensors in a robot hand[37]. Piezo-electric elements can also be used for low-force actuation, problems of hysteresis, creep and non-linearity with large signals being overcome by feedback from silicon strain gauges bonded to the crystal[38].

Where, as in a robot, many strain gauges are required to be in close proximity, it has been found advantageous to use pulse excitation, since by this means not only are the signal levels increased some thousand times, but the signals from different gauges are separated in time and so proximity interference is minimised[45-47].

Velocity and acceleration feedback

Where electric d.c. motors are used for actuation, there is no general problem in obtaining velocity signals for purposes of feedback in the movement servo-control systems. If the magnetic field of an electric motor is maintained constant in value, then the armature voltage contains a major term which is proportional to the rotational speed of the motor.

There are, however, additional terms contained in the armature voltage which can introduce errors if the armature voltage is to be used to indicate speed. First of these is the armature voltage drop caused by the armature current which flows in the resistive armature winding. Secondly, there is the voltage drop caused by the brush gear of the motor. Thirdly, there is the ripple and electrical noise generated by the movement of the armature, by the brush gear and by the rotational variations of the reluctance of the magnetic circuit.

Fortunately, for many purposes these errors can be avoided, and it is then possible to make use of the voltage available at the motor armature for velocity feedback purposes. This is particularly true if a constant-current form of armature supply, using, for example, the Boucherot circuit[27], is adopted.

Tachometer generators to provide speed feedback signals are unlikely to be fitted to the controls of mobile robots because of the added complication, weight and cost involved. In general, a close control of rotational speed is much less necessary than control of limb position in the robot.

Acceleration feedback can be a valuable feature of a control system. However, accelerometers are unlikely to be used where feedback of acceleration signals is required in the mobile robot, because of the additional weight and complication involved. Instead, acceleration signals can be obtained by the differentiation of velocity signals. This course is not free of problems, caused by, for example, tachometer brush ripple, but these problems and possible solutions are well-known. Robot devices for very accurate long-range navigation require very accurate measuring devices for velocity and acceleration determination, though the additional expense is justified in such cases and precision devices are then used.

REFERENCES

1. Houk, J. C., et al., 'Frequency Response of a Spindle Receptor', *MIT Res. Lab. Electron. Q. Prog. Rep.*, **67**, October, 223 (1962).
2. Hammond, P. H., 'Living Control Systems', *Electron. & Power*, **13**, September, 338 (1967).
3. Marko, H., 'Information Theory and Cybernetics', *IEEE Spectrum*, **4**, November, 75 (1967).
4. Worden, F. G., 'Hearing and the Neural Detection of Acoustic Patterns', *Behav. Sci.*, **16**, 20 (1971).
5. Blinkov, S. M. and Glezer, I. I., *The Human Brain in Figures and Tables*, Plenum (1968).
6. Tomovic, R. and Boni, G., 'An Adaptive Artificial Hand', *Trans. IRE*, **AC7**, April, 3 (1962).
7. Giles, A. F., *Electronic Sensing Devices*, Newnes (1966).
8. Anon., 'An Electronic Nose', *Electron. Eng.*, **20**, November, 367 (1948).
9. Lawrence, S. J., et al., 'The Solid Electrolyte Oxygen Sensor', *Proc. Joint Autom. Control Conf. AACC, Colorado, 1969*, 749.
10. Shaw, E. N., 'Heat Control in Electronic Equipment', *Electron. Eng.*, **29**, January/February/March, 13, 65, 115 (1957).
11. Shaw, E. N., 'Liquid Cooling of Electronic Equipment', *Electron. Eng.*, **30**, September, 516 (1958).
12. Batcher, R. R. and Moulic, W., *The Electronic Control Handbook*, 79, Caldwell-Clements (1946).
13. Bryce, C. H., and Hole, V. H. R., 'Measurement and Control by Thermistors', *Ind. Electron.*, **5**, July/August, 294, 358.
14. Scarr, R. W. A. and Setterington, R. A., 'Thermistors, Their Theory, Manufacture and Application', *Proc. IEE*, **107 B**, September, 395 (1960).
15. McCann, M. R., 'Calculation of V–I Curves for NTC Thermistors', *Electron. Eng.*, **39**, June/July, 346, 426 (1967).
16. Young, John F., 'A Method of Precise Position Indication', *Control*, **11**, December, 588 (1967).
17. Van Santen, G. W., *Electronic Weighing and Process Control*, Macmillan (1967).
18. Andrews, H. I., 'Electrical Weighing', *Proc. IEE*, **97**, pt 1, May, 98 (1950).
19. Woodcock, F. J., 'Some Electrical Methods of Measuring Mechanical Quantities', *JIEE*, **97**, pt 1, July, 136 (1950).
20. Tiffany, A. and Wood, J., 'Precision Strain Gauge Techniques', *Electron. Eng.*, **30**, September, 528 (1958).
21. Chandler, R. L. and Dent, E. J., 'Temperature Compensated Strain Gauges', *Electron. Eng.*, **32**, July, 414 (1960).
22. Crawford, J. C., 'A Piezoelectric Field Effect Strain Gauge', *Exp. Mech.*, **11**, April, 145 (1971).
23. Elphee, E. K., 'Tachometry in Industry', *Ind. Electron.*, **1**, November, 713 (1963).
24. Baker D., et al. (eds.), *Design Technology*, 486, Prentice-Hall (1970).
25. Winton, H. J. and Linebarger, R. N., 'Computer Simulation of Human Temperature Control', *Simulation*, **15**, October, 213 (1970).
26. Young, John F., *Applied Electronics*, 156, Iliffe (1968).
27. Young, John F., 'The Boucherot Effect', *Wireless Wld*, **68**, August, 391 1962).
28. McCormick, E. J., *Human Factors Engineering*, McGraw-Hill (1964).
29. Berlincourt, D. A., 'Piezo-Electric Transducers', *Electro-Technology, N.Y.*, **85**, January, 39 (1970).

30. Kadlec, C., 'The Piezo-Junction Transducer', *Electro-Technology, N.Y.*, **85**, January, 39 (1970).
31. Anon., 'A Transistor Temperature Sensor', *Ind. Electron.*, **5**, August, 357 (1967).
32. Wright, R. H., et al., 'Olfactory Coding', *Nature, Lond.*, **216**, 404 (1967).
33. Wright, R. H., 'How Animals Distinguish Odours', *Sci. J.*, **4**, July, 57 (1968).
34. Kidd, S., 'Communicating through the Skin', *New Scient.*, **30**, April 14, 82 (1966).
35. Heist, H. E., 'Can Flavour Be Controlled Automatically?', *New Scient.*, **32**, November 10, 298 (1966).
36. Keidel, W. D., 'Tuning between Central Auditory Pathways and the Ear', *Trans. IEEE*, **MIL7**, April/July, 131 (1963).
37. Kinoshita, G., et al., 'Pattern Recognition by an Artificial Tactile Sense', *Proc. 2nd Int. Joint Conf. Artificial Intelligence, London, September, 1971* (Br. Computer Soc.), 376.
38. Anon., 'Piezo-Electric Actuators for Precision Movements', *Electron. Eng.*, **44**, June, 30 (1972).
39. Anon., 'Gas-Sensing Semiconductor', *Electron. Eng.*, **43**, August, 29 (1971).
40. Ranson, J. B., 'Real-Time Oxygen Measurement for Combustion Control', *Control & Instrum.*, **4**, September, 46 (1972).
41. Ruskin, R. E. (ed.), *Humidity and Moisture*, Vol. 1, 219, Reinhold (1965).
42. Thoma, P. E., 'A Resistance Humidity Sensing Transducer', *Ind. Engng Chem. Ind (Int.) Edn*, **19**, May, 30 (1972).
43. Hole, V. H. R., 'Thermistor Temperature Measuring Bridge Circuits', *Electron. Components*, November 12, (1971); January 14, (1972).
44. Wlodarski, W., 'Possibility of the Application of the Zener Diode as a Hydrostatic Pressure Gauge', *Electron. Lett.*, **6**, February 5, 64 (1970).
45. Yates, J. G., 'Pulse Excitation of Impedance Bridges', *Nature, Lond.*, **163**, 132 (1949).
46. Prowse, W. A. and Laverick, E., 'Differential Transformer Bridge Operated by Square Waves', *Nature, Lond.*, **163**, 571 (1949).
47. Yates, J. G., et al., 'Multi-Channel Measurement of Physical Effects by Confluent Pulse Technique, with Particular Reference to the Analysis of Strain', *Proc. IEE*, **98**, pt 2, April, 109 (1951).
48. Anon., 'Transducer Built and Priced Like I.C.', *Electronics*, **45**, October 9, 41 (1972).
49. Slocombe, J. W., et al., 'Capacitive Moisture Measurement', *Instr. Contr. Systems*, **42**, March, 131 (1969).
50. Shuhei, A. and Gen-Ichiro, K., 'A Pattern-Classification by the Time-Varying Threshold Method-Application of the Method to Visual and Tactile Senses', *Proc. Int. Joint Conf. Artificial Intelligence, May, 1969*, 417.
51. Kaufman, A. B., 'Velocity and Acceleration Transducers', *Instr. Contr. Systems*, **44**, April, 115 (1971).
52. Kakikura, M., et al., 'On the Control of an Industrial Robot with Tactile Sensors', *Trans. Soc. Instr. Contr. Eng., Japan*, **7**, 31 (1971).
53. McCulloch, W. S., 'Why the Mind is in the Head', in: *Embodiments of Mind*, 72, M.I.T. (1965).
54. Borisova, L. F. and Lenshina, L. K., 'Analysis of Mathematical Model for Mutual Inhibition Networks', *Autom. Rem. Control*, **33**, December, 1189 (1972).
55. Fan, L. T., et al., 'Review on Mathematical Models of the Human Thermal System', *Trans. IEEE*, **BME18**, May, 218 (1971).

56. Shitzer, A., 'Addendum to Ref. 55', *Trans. IEEE*, **BME20,** January, 65 (1973).
57. Wise, K. D. and Angell, J. B , 'An I.C. Piezoresistive Pressure Sensor for Biomedical Instrumentation', *Trans. I.EEE*, **BM20,** March, 101 (1973).
58. Crochetiere, W. J., 'Engineering Design for the Disabled', *Trans. IEEE*, **E16,** February, 59 (1973).
59. Young, John F., *Cybernetic Engineering*, Butterworths (1973).

3

The muscles of the robot

The dynamic performance of muscle[1-5, 41]

In some applications of robots it will be necessary to simulate the performance of human muscle as closely as possible. Consequently, it is of interest to consider the nature of human muscle action.

The relationship between the force F exerted by the muscle on the load being moved and the velocity V of movement can be expressed by the relationship

$$F = sF_{max} \times \frac{1 - V/V_{max}}{1 + k\ V/V_{max}}$$

Here s is fractional stimulation = actual stimulation/maximum stimulation; F_{max} is maximum force; V_{max} is maximum velocity; and k is a constant.

The expression leads to the graphical relationship of *Figure 3.1*, where k has been taken as equal to 5.

A typical curve of torque against speed for an average elbow joint[6, 33, 34] of an adult human male is shown in *Figure 3.2*. Prosthetic devices usually give a performance far below this curve—for example, as shown dotted. However, although this performance is far below the maximum curve, it is not far from the performance which many would consider as normal, as shown. Note that a curve of constant power is a rectangular hyperbola on these axes, so that it can be seen that an output of 20 W is adequate for most average tasks.

The practical curve obtained for a human is somewhat below the curve for constant power output, since constant power corresponds to a constant product of force × velocity.

The Muscles of the Robot 43

An alternative way of expressing the relationship obtained between force and velocity is

$$(F+a)(V+b) = \text{constant} = b(F_0+a) = a(V_0+b)$$

where F and V are force and velocity; a and b are constants; F_0 is the force when $V = 0$; and V_0 is the velocity when $F = 0$. The constant a is approximately equal to $F_0/3$. The maximum value of the power is given by

$$\frac{F}{a} = \left(\frac{F_0}{a}+1\right)^{\frac{1}{2}} - 1$$

as can be seen by differentiation and equation to zero.

Kremer[44] has reviewed the knowledge of human strength, and he has shown how the endurance time falls rapidly as the muscle force

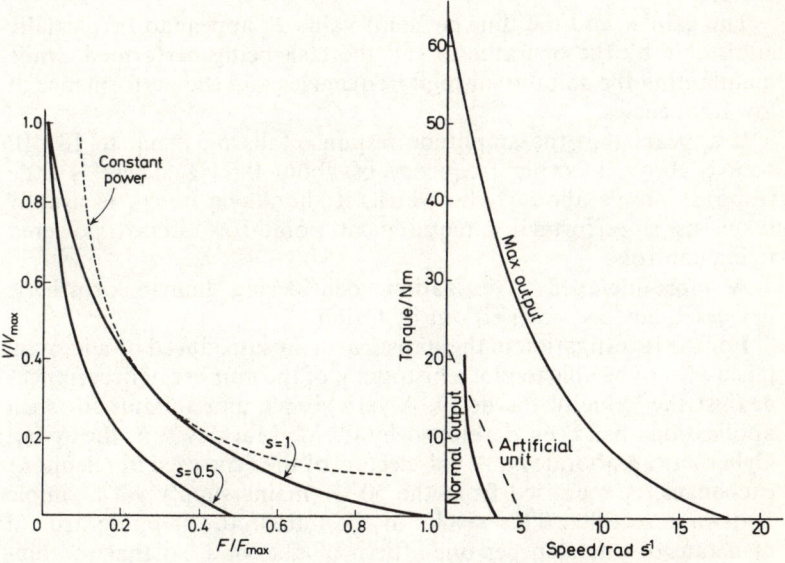

Figure 3.1 Relationship of force exerted by muscles

Figure 3.2 Torque/speed curve for average elbow joint

requirement is increased. The nonlinear relationship between static tension and muscle length has been discussed by Inman and Ralston[47].

Human transfer function[7-10, 23, 24, 26, 27]

There is, of course, no basic reason why the operation of a robot should approximate to that of a human. However, in certain cases— for example, when it is necessary to operate several different machines designed originally for use by humans—it is desirable that the transfer function of the robot controls should approach that of the human being replaced. This is true of the general-purpose robot for use in the human environment.

Experiments indicate that the transfer function of a human operator is typically given by

$$\frac{\text{output}}{\text{input}} = K \frac{(1+sT_2)}{(1+sT_1)(1+sT_3)} \exp(-sT_d)$$

Here the delay time is $T_d \approx 0.2$ s (human reaction time); $T_1 \approx 10$ s (l.f. stabilising); $T_2 \approx 0.5$ s (h.f. stabilising); $T_3 \approx 0.1$ s (inertia lag); and the gain $K \approx 2$–100. However, all of these values are very variable*.

The gain K and the time constant value T_3 appear to be partially adjustable by the operator to suit the task being performed, while maintaining the stability at high frequencies and the performance at low frequencies.

It appears that the amplitude response falls off at up to 18 dB/octave above a corner frequency of about 0.6 Hz. There is little response above about 3 Hz, which, it should be noted, is usually taken as a performance requirement point for aircraft powered flying controls.

A more detailed investigation, considering human prosthetic devices[29], has been carried out by Orloff.

For the investigation of the statistical delay introduced by a human, it is useful to be able to plot a histogram of the number of occurrences against the value of the delay. A very simple unit suitable for such applications has been developed by C. M. Marklew[59] in the Aston Cybernetics Laboratory. A uniselector, of the type used in telephone exchanges, is energised from the 50 Hz mains supply via a simple half-wave rectifier. This results in the uniselector stepping around at a rate of one step per one-fiftieth of a second, so that the uniselector performs one revolution each second. In this way the uniselector can be used as a continuously-rotating 50-way switch. The input to be observed is applied to the rotor of the uniselector.

---

* However, the writer has noted that although published values of the time constants vary over at least a 100 : 1 range, the ratios of, for example, T_1/T_3 are more consistent, within little more than a 2 : 1 range.

A total of 50 capacitors is connected to the switch contacts. Consequently, each of the capacitors corresponds to a different value of delay time. If the starting of the uniselector causes the initiation of a process, such as the stimulation of a nerve cell, then the moving contact can be used to store the result on the various capacitors, each capacitor corresponding to one point in time up to 1 s from the stimulation. The same moving contact can then extract the information stored on the capacitors for later display—for example, on a storage type of oscilloscope or for photography.

It is perhaps as well to note that much of the experimental work using animal and human muscles has assumed operation in the linear region. In emergencies, when operation is in the nonlinear region, very high speeds of movement can be obtained. Taylor has suggested that restriction of analyses to the small-scale linear approximation region might now be delaying progress[49].

Performance requirements of hand and arm[50-56, 60]

It has been suggested that the minimum requirements of a human-type arm are:

1. Flexion, extension, abduction and adduction at the shoulder (i.e. movement in two planes).
2. Rotation, either at the shoulder or, for convenience, in the upper arm.
3. Movement of the elbow.
4. Bending of the wrist.
5. Rotation of the wrist.
6. Prehension (or grasping).

It will be advantageous if the limbs of some robots can move in a natural fashion, not only from the point of view of appearance to humans but also from considerations of control and co-ordination in the normal human environment. It will therefore be of interest to examine the results of investigations into the performance of human limbs, in order to obtain a standard for comparison for the performance of robot limbs.

The peak angular velocity of human arm joints is of the order of 2 rad/s, lasting for a period of 1 s, while finger movements and forearm rotations are from two to three times faster than this. Lifting a mass of 5 kg has the effect of halving the peak velocity. The peak rate of elbow movement is rather less than 17 rad/s and that of forearm rotation about twice as much.

It is interesting to note that young children sometimes even start a movement in the wrong direction, in addition to overshooting the final position. We must not be surprised to encounter such actions in the untrained robot.

A static elbow-torque requirement of 70 Nm with an unloaded full range of movement repeatable at a rate of 3 Hz have been suggested. Figures such as those presented here give a preliminary idea of the engineering requirements of the control system for a robot arm capable of a quasi-human performance.

The deformability of human fingertips is a most important feature which helps the fingers to perform their tasks. On the contrary, however, the relative nondeformability of the fingernail gives a great advantage when small objects have to be picked up. In order to signal information back to the central nervous system, the leverage of the fingernails is also very important. A robot without soft-surfaced fingers and fitted with rigid fingernails is probably at a disadvantage when compared with a human being for many tasks.

For heavier tasks, such as wielding a hammer or a screwdriver, it should be noted that the human commonly uses one hand for power and the other for fine guidance. From this viewpoint it is desirable that any robot should have at least two hands.

From the point of view of standardisation and low cost, it is desirable that a robot should have general-purpose hands. However, there is no reason why special-purpose hands should not be fitted so that tasks which are difficult or impossible to a human can be performed. It is also advantageous for a robot to be fitted with power sockets, so that it can power portable electric tools which it is using. Like all other parts of a robot, the hands should be easily replaceable.

A basic problem in the dynamics of a robot arm is imposed by the relationship between the positioning accuracy required and the positioning time necessary. Since any system is likely to be to some extent oscillatory, the natural frequency must be made as high as possible in order to achieve rapid positioning. Now in the case of a balanced system the weight of the robot body is balanced against the weight being lifted. The required positioning time in such a case increases as the accuracy of positioning is increased. Dampers and dashpots can be used to reduce the oscillation, though it is possible to provide electrical adjustment if electronic servo-control is incorporated[11].

It has been estimated[35] that the number of daily movements of a human arm prosthesis is about 2000, or 150 000 cycles for a period between servicings of 3 months. The robot arm should be capable of at least this number of cycles between servicings, and the aim should be to achieve a greatly increased figure.

Reduction to essentials

Essentially, a hand can only move in three different dimensions, and therefore only one control in each of the axes, x,y,z, should be required. An alternative approach is to consider the control of radius and of two angles. Again only three independent controls are required. Thus, if the forces imparted by the actuators are sufficiently in excess of all needs in all directions, there is basically only a need for three separate control systems on a non-realistic robot arm.

However, such an approach involves an excessive over-simplification in that it ignores the fact that a human elbow, and to some extent a human wrist, can be bent in order to enable the hand to reach around obstacles. This greatly increases the range of tasks which can be performed by a human.

Some economy would be obtained by the reduction of some simple robots to essentials. This has in fact been done with some of the early machine loading robots. However, the resulting loss of control flexibility severly restricts the value of such robots in the general human environment. From an economic point of view, there will be a greater market for the general-purpose robot than there is for the very much more restricted special-purpose device.

Feedback control is likely to be essential in most applications. This can include the usual feedback of output, as in a servo loop. In a robot as in an animal system, it can also include control by completely opposing sets of muscles, each with its own control system which also inhibits one muscle as it actuates the opposing muscle.

In addition to the three independent and bidirectional forces and motions of which a robot hand must be capable, it is also desirable to consider the necessity for three independent and bidirectional couples and rotations. In general, these can be provided by a form of wrist action.

Bottomley[30] has shown how, if the distance between elbow and hand is equal to the distance between elbow and shoulder, it can be ensured that the hand always moves radially towards or away from the shoulder. A toothed pulley is fitted at the shoulder and another of one-half the diameter is fitted to the elbow and locked to the forearm. The two pulleys are connected by a toothed belt or bicycle-type chain. The elbow then sweeps through twice the angle traversed by the shoulder, while the hand moves radially. Bottomley also describes the ingenious Steeper locking device which locks the angle of, for example, an elbow unless there is a slow pull or relaxation of the operating cable.

Grasping movements of fingers[12-16, 45]

It is necessary that the general-purpose robot hand should be capable of reproducing as many as possible of the normal human hand and finger movements. The possible grasping movements which can be made by the human fingers can be categorised as follows:

1. The *palm* grip of a reasonably sized object, with the thumb opposed to the finger tip.
2. The *tip* grip, of a thin object, with the thumb and finger tips opposed.
3. The *pen* grip, used for writing or cutting.
4. The *lateral* grip, between the tip of the thumb and the side of the forefinger.
5. The *rod* grip, with the fingers and the thumb opposingly curled around a rod.
6. The *ball* grip, with all of the fingers and the thumb fairly evenly spaced around the ball.
7. The *hook* grip, with one or more fingers hooked into a handle or a cavity.

There is a basic difference between pinching (as with the tip grip) and grasping (as with the rod grip). This is that in the rod grip there is a delay of the movement of the thumb compared with the movement of the fingers. Tomovic and Boni[15] have proposed a most ingenious solution for this requirement. The two cables which operate the thumb and the fingers are simply taken to different angular positions on the same pulley. Then with one direction of rotation (for the tip grip), the thumb moves at the same time as the fingers. With the opposite direction of rotation, the thumb movement is delayed as required for rod grip. They use a 4 W motor with 100:1 gearing to drive this hand.

In early work on remote controlled manipulators it was found that only two opposed fingers, having a vice-action grip, were sufficient for many applications. In early work wrap-around types of tong were experimented with, but the parallel-jaw type was found to be superior[17, 58]. However, in future work on human-controlled manipulators it is likely that a more complex action will be adopted, and of course such work is already in progress on human prosthetic devices (e.g. the Rakic hand[32]).

An ingenious type of gripping device has been used in some applications of the Versatran industrial robot. Fixed opposed fingers have each been fitted with an inflatable rubber or plastic pad. When both pads are inflated, an irregularly shaped object can be gripped securely.

It is very likely that such a non-human method of gripping will be more widely applied in the future for robot applications.

Ring[28] has added a detector of sideways motion or slip to control the grip exerted by a robot hand. The gripping force can be related to the weight of the object being held if the coefficient of friction between the hand and the object can be assumed to be a constant. A time response of one-tenth of a second has been achieved with such a device.

Thring[31] has described a form of seven-fingered hand with one degree of freedom, which can grip a wide variety of objects. The seven fingers are relatively independent and they oppose a fixed 'thumb' but unlike the cosmetic Rakic hand[32], which also has relatively independent fingers, there is no attempt to simulate the appearance of a human hand. With a robot this is unnecessary.

If only one control, opposed by a spring, is to be fitted to the fingers of a hand, it has been found in work on prosthetics that the normally open hand is to be preferred to the normally closed type[45]. If a two-position thumb is opposed by fingers which can move through 40 mm. objects up to 80 mm. diameter can be grasped, though 90% of the time an opening of less than 40 mm is required. The grasping force should be at least 70 N. A 'three-jaw chuck' type of hand, with two fingers opposing the thumb, is most successful, and in prosthetic use the other two fingers have sometimes been latex dummies.

The high coefficient of friction of human skin is a big advantage where objects have to be gripped and picked up. The coefficient of friction of human skin is about 3, and it falls little even under water. It is possible to obtain some plastics and rubbers with comparable figures. Various rubbers, leathers and plastics have been used for prosthetic 'skin', though the PVC materials seem to have advantages[46]. Robot skin should be waterproof and as resistant to other liquids and to weathering as possible.

Godden[36] has pointed out the big advantages of a multi-point grip in supporting heavy objects. When slip is about to occur, it begins first to appear at the edges of the contact area[42].

It is difficult to envisage at present any skin covering for the robot which will be anywhere near as versatile as human skin[37-40, 43, 57], which provides sensors such as touch and heat sensors, provides liquid rejection and yet also provides for evaporative cooling and is continuously self-renewing and self-repairing. Early robots will be metallic, perhaps with a form of skin on the fingers.

Recently a new synthetic plastic or 'polyion complex' has been announced for use mainly as a human tissue substitute in medicine,

50 The Muscles of the Robot

surgery and pharmacy[48]. This is a polystyrene derivative, which is hydrophilic, or 'water-loving', and can be produced as films or coatings for various plastics, metals and glasses. It is expected that such a substance will have many medical applications, but if it can be produced at low cost then it will be very suitable for use as a coating for the mobile robot also.

The lifting action in the robot[18, 61]

One of the major tasks of the general-purpose robot will be to pick up various loads prior to their transport. It is therefore necessary to consider the weight which can be lifted. The basic problem is that, in general, the weight must be lifted along a line which does not pass

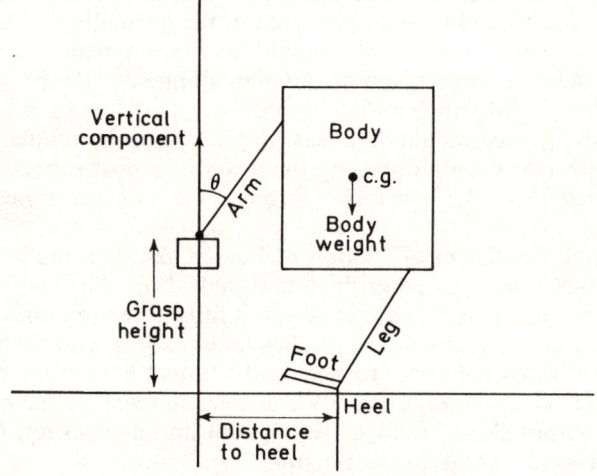

Figure 3.3 Lifting action of the robot

through the base or feet of the robot and which does not pass through the robot centre of gravity (c.g.). Similar problems are, of course, encountered with any lifting device, such as a crane. However, it is of interest to consider briefly the action of the humanoid robot in lifting.

There are two general principles which can be laid down in the form of general lifting instructions to the robot. Firstly, the weight to be lifted should be grasped near to its base. Secondly, the weight to be lifted should be positioned as near as possible to the base or feet of the robot.

In effect, the robot body should be used as a lever, the weight of the body being balanced against the weight being lifted. The arms should be as vertical as possible. The fulcrum of the lifting leverage is the heel, or the point on the base of the robot which is furthest from the load.

The relationship between the various forces has been expressed as follows:

vertical component of lifting force

$$= \frac{\text{constant } k \times \text{body weight (acting through c.g.)}}{(\text{distance from weight lifted to heel} + \text{grasp height above ground} \times \tan \theta) - \text{const. } b}$$

where θ is the angle to the vertical of the lifting arm, as shown in *Figure 3.3*.

The relationship between the weight of an object being lifted by a human being and the height to which the human can lift it without injury or fatigue has been investigated. A fairly complex curve of weight against height is produced. However, in order to ensure that a robot is capable of carrying out human tasks, it is only necessary to approximate to the experimentally determined human characteristics.

The present writer's suggested approximation is that the relationship between the maximum weight F which can be lifted, expressed in kilograms, and the height H in metres to which it can be lifted by a human is expressed by

$$F^2 + 1600 H^2 = 3600$$

This relationship gives a maximum value of 60 kg which can be just lifted from the floor, and a maximum height to which any weight can be lifted of 1.5 m, the latter being clearly an underestimate for any normal human. The simplicity of the relationship makes it an attractive one to set as a minimum robot requirement, however.

Introduction of backlash[19, 20]

It has been suggested that the characteristic of movement of an artificial limb, when plotted against the average signal from a nerve, should not simply be a true straight line. Instead, a dual form of characteristic response is required. There should be a slow rate of change of force causing movement, following a true straight line through the origin of the signal–force curve. In addition, there should be a rapid rate of movement if the signal is greater in value than a certain value, so introducing a second line on the signal–force plane which does not in this case pass through the origin.

When both directions of movement are considered, this procedure leads to the form of characteristic shown in *Figure 3.4*. A simple biased-diode backlash arrangement in the electronic control circuitry can easily introduce this form of characteristic to the control servo. However, it has been found advantageous in prosthetic

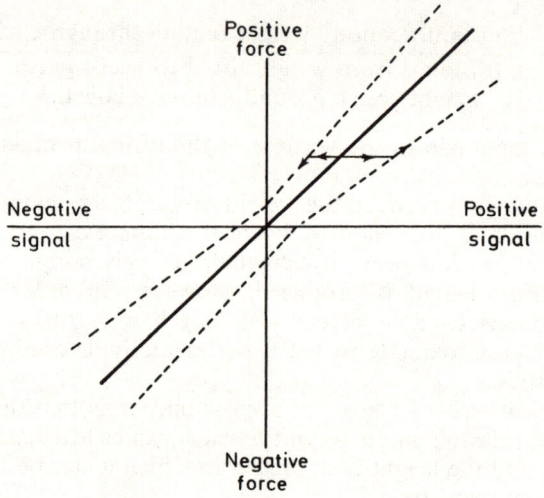

Figure 3.4   Backlash characteristic

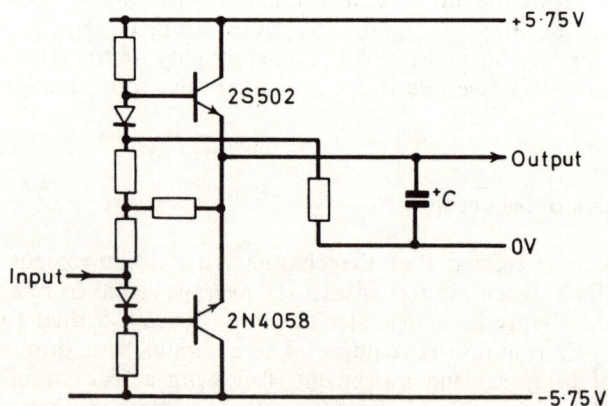

Figure 3.5   Transistor backlash circuit

devices to incorporate more complex backlash circuitry utilising transistors.

A typical transistor backlash circuit [21, 22] is shown in *Figure 3.5*. With this arrangement the output voltage can only change slowly in response to input voltage changes of small amplitude, since there is an effective *R–C* smoothing circuit in the input-output path. However, if there is a sudden change of more than about ± 30% in the input voltage, then the capacitor is charged or discharged rapidly by the transistors. One valuable feature of this circuit is the extremely low value of steady power consumption.

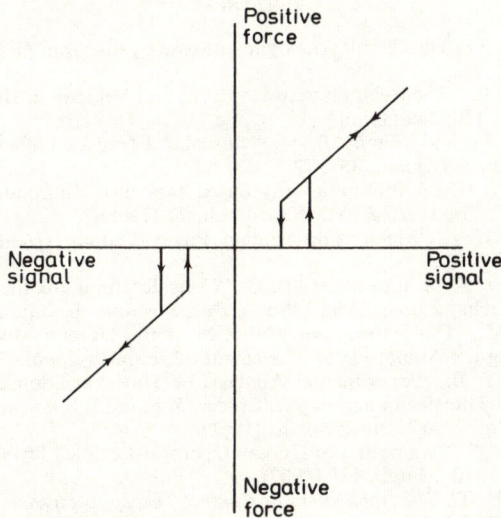

Figure 3.6 *Force/input signal characteristic of a typical prosthetic device*

The form of backlash circuit shown in *Figure 3.5* has been used in prosthetics to avoid the effects of fluctuations of the input from the human nerve fibres. Other forms of backlash and hysteresis have been used in prosthetics in order to achieve battery economy, as will be described later. As an example, the over-all output force–input signal characteristic of a typical prosthetic device is shown in *Figure 3.6*. There is no response until the input exceeds a certain value, when the response is linear. When the input change reverses in direction, there is a hysteresis effect, designed to avoid instability. A delay device is included in the backlash arrangements, so that the response is linear to very rapid changes of input. This permits rapid forward and reverse movements.

Backlash arrangements can sometimes be used to help to limit the effects of interference such as mains supply-frequency pick-up. However, it is often desirable in portable equipment such as a robot to install plenty of notch filters which reject signals at supply frequency, since these interfering signals can be most troublesome in a mobile device.

The problems of wear and its reduction and prevention in the joints of a robot are clearly important points. Here, once more, the robotics engineer can learn much from the experiences of engineers who have worked on artificial limbs[25].

REFERENCES

1. Wilkie, D. R., 'The Circuit Analogue of Muscle', *Electron. Eng.*, **22**, October, 435 (1950).
2. Wilkie, D. R., 'The Relation between Force and Velocity in Human Muscle' *J. Physiol.*, **110**, 249 (1950).
3. Fenn, W. O. and Marsh, B. S., 'Muscular Force at Different Speeds of Shortening', *J. Physiol.*, **85**, 277 (1935).
4. Vickers, W. H., 'A Physiologically Based Model of Neuromuscular System Dynamics', *Trans. IEEE*, **MMS9**, March, 21 (1968).
5. Harrison, J. Y., 'Maximising Human Power Output', *Hum. Factors*, **12**, June, 315 (1970).
6. Livingstone, S. M. and Styles, B. C., 'Considerations for the Design of an Electro-Mechanical Artificial Elbow', *Des. Electron.*, **5**, July, 20 (1968).
7. Fitts, P. M., 'The Information Capacity of the Human Motor System in Controlling the Amplitude of Movement', *J. Exp. Psychol.*, **47**, 381 (1956).
8. Sheridan, T. B., 'Experimental Analysis of Time-Variation of the Human Operator's Transfer Function', in: Coales, J. F. (ed.), *Automatic and Remote Control*, Vol. 2, 629, Butterworths (1961).
9. Garner, K. C., 'Evaluation of Human Operator Coupled Dynamic Systems', *Ergonomics*, **10**, March, 125 (1967).
10. Milhorn, H. T., *The Application of Control Theory to Physiological Systems*, Saunders (1966).
11. Van Santen, G. W., *Electronic Weighing and Process Control*, Macmillan (1967).
12. Schlesinger, G., 'Die Mitarbeit des Ingenieurs bei der Durchbildung des Ersatzglieder', *Ver. Dtsch. Ingen.*, Berlin, **61**, 6 (1917).
13. Tomovic, R., 'The Human Hand as a Feedback System', in: Coales, J. F. (ed.), *Automatic and Remote Control*, Vol. 2, 624, Butterworths (1961).
14. Kobrinski, A. E., *et al.*, 'Problems of Bio-Electric Control', in: Coales, J. F. (ed.), *Automatic and Remote Control*, Vol. 2, 619, Butterworths (1961).
15. Tomovic, R. and Boni, G., 'An Adaptive Artificial Hand', *Trans. IRE*, **AC7**, April, 3 (1962).
16. Horn, W. G., 'Muscle Voltage Moves Artificial Hand', *Electronics*, **36**, 32 (1963).
17. Goertz, R., 'Manipulator Systems Development at ANL', *Proc. 12th Conf. Remote Systems Technology, 1964*.
18. Whitney, R. J., 'The Strength of the Lifting Action in Man', *Ergonomics*, **1**, February, 101 (1958).
19. Bottomley, A. H. and Cowell, T. K., 'An Artificial Hand Controlled by the Nerves', *New Scient.*, **21**, March 12, 668 (1964).

20. Bottomley, A., et al., 'Muscle Substitutes and Myo-Electric Control', *J. Br. IRE*, **26**, 439 (1963).
21. Fyson, J., et al., 'Design Considerations in a Myoelectric Hand Prosthesis', *Proc. IEE*, **116**, February, 281 (1969).
22. Allison, R., et al., 'British Patents 13371/65, 13661/65, 15204/65, 45639/65.
23. Sutton, C. G., 'The Human Operator in a Control System', *Br. Comm. Electron.*, **4**, December, 744 (1957).
24. Shearer, J. L., et al., 'Dynamics of Human Operators', *Control*, **4**, January, 113 (1961).
25. 'Lubrication and Wear in Living and Artificial Human Joints', *Proc. Instn Mech. Engrs*, **181**, pt 3J (1967).
26. Mitchell, M. B., 'Systems Analysis—The Human Element', *Electro-Technology, N.Y.*, **77**, April, 59 (1966).
27. Costello, R. G. and Higgins, T. J., 'Bibliog. on Human-Operator Modelling', *Trans. IEEE*, **HFE7**, December, 174 (1966).
28. Ring, N. D., 'A Study of Prehension', M.Sc. Thesis, Engng Dept., University of Cambridge (1966).
29. Kennaway, A. (Chairman), *Symposium on Powered Prostheses, Roehampton, October, 1965* (Ministry of Health).
30. Bottomley, A. H., 'An Approach to a Powered Arm with Co-ordinate Control', *Proc. Instn Mech. Engrs*, **183**, pt 3J, November, 82 (1968).
31. Thring, M. W., 'The Next Thirty Years in Engineering', *Advan. Sci.*, **24**, September, 99 (1967).
32. Rakic, M., 'The Belgrade Hand Prosthesis', *Proc. Instn Mech. Engrs*, **183**, pt 3J, November, 60 (1968).
33. McWilliam, R., 'Some Characteristics of Normal Movement in the Upper Limb', *Proc. Symp. Powered Prostheses, Roehampton, 1965*.
34. Livingstone, S. M. and Crecraft, D. I., 'Design of an Artificial Elbow: An Electromechanical Solution', *Proc. Instn Mech. Engrs*, **183**, pt 3J, November, 32, (1968).
35. Montgomery, S. R., 'Design of an Experimental Arm Prosthesis: Engineering Aspects', *Proc. Instn Mech. Engrs*, **183**, pt 3J, November, 68 (1968).
36. Godden, A. K., 'Some Factors in the Design of an Adaptive Artificial Hand', *Proc. Instn Mech. Engrs*, **183**, pt 3J, November, 50 (1968).
37. Montagna, W., *The Structure and Function of the Skin*, Academic Press (1962).
38. Anon., 'An Engineering Study of Human Skin', *Engineering, Lond.*, **199**, 363 (1965).
39. Sinclair, D., *Cutaneous Sensation*, Oxford University Press (1967).
40. Sternbach, R. A., *Pain*, Academic Press (1968).
41. Pringle, J. W. S., *Models of Muscle* (Symposia of the Society for Experimental Biology, No. 14), 41, Cambridge University Press (1960).
42. Nightingale, J. M. and Todd, R. W., 'Adaptive Control of a Multi-Degree of Freedom Hand Prosthesis', *Conf. Hum. Locomotor Eng., Sussex, September 1971* (Instn Mech. Engrs), 249.
43. Ridge, M. D. and Wright, V., 'A Rheological Study of Skin', in: Kenedi, R. M. (ed.), *Biomechanics and Related Bio-Engineering Topics*, 165, Pergamon (1965).
44. Kroemer, K. H. E., 'Human Strength', *Hum. Factors*, **12**, June, 297 (1970).
45. Fletcher, M. J., 'New Developments in Hands and Hooks', in: Klopsteg, P. E. and Wilson, P. D. (eds.), *Human Limbs and Their Substitutes*, 222, McGraw-Hill (1954).

46. Leonard, F. and Milton, C. L., 'Cosmetic Gloves', in: Klopsteg, P. E. and Wilson, P. D. (eds.), *Human Limbs and Their Substitutes*, 239, McGraw-Hill (1954).
47. Inman, V. T. and Ralston, H. J., 'The Mechanics of Voluntary Muscle', in: Klopsteg, P. E. and Wilson, P. D. (eds), *Human Limbs and Their Substitutes*, 296, McGraw-Hill (1954).
48. Anon., 'Water-Loving Plastic Seen as Tissue Replacement', *Ind. Res.*, **13**, October, 21 (1971).
49. Taylor, A., 'The Limitations Imposed on the Muscle Servo by the Nature of its Components', *IEE Meet. Control Muscle Contraction Animal Movement, May 8, 1972*.
50. Napier, J., 'The Evolution of the Hand', *Sci. Am.*, **207**, December, 56 (1962).
51. Daniels, G. S. and Hertzberg, H.T.E., 'Applied Anthropometry of the Hand', *Am. J. Phys. Anthropol.*, **10**, 209 (1952).
52. Tomovic, R., 'Human Hand as a Feedback System', in: Coales, J. F. (ed.), *Automatic and Remote Control*, Vol. 2, 622, Butterworths (1961).
53. Yamashita, T. and Mori, M., 'Engineering Approaches to Function of Fingers', *Rep. Inst. Ind. Sci., Tokyo Univ.*, **13**, November, 60 (1963).
54. Long, C., 'Intrinsic-Extrinsic Muscle Control of the Fingers', *J. Bone Jt. Surg.*, **50A**, July, 973 (1968).
55. Merton, P. A., 'How We Control the Contraction of our Muscles', *Sci. Am.*, **226**, May, 30 (1972).
56. Neilson, P. D., 'Speed of Response or Bandwidth of Voluntary System Controlling Elbow Position in Intact Man', *Med. Biol. Eng.*, **10**, July, 450 (1972).
57. Von Bally, K., *et al.*, 'Development of a Bio-Compatible Rubber for Implantable Artificial Organs or Instrumentation', *J. Audio Eng. Soc.*, **18**, December, 692 (1970).
58. Aldred, J. K., *et al.*, 'Mechanical Upgrading of Industrial Robots', *Proc. 19th Conf. Remote Systems Technology, Miami, October, 1971*, 69.
59. Marklew, C. M., 'Pulse Delay Unit and Histogram Plotter', University of Aston (1970).
60. Martin, J. B., and Chaffin, D. B., 'Biomechanical Computerized Simulation of Human Strength in Sagittal Plane Activities', *Trans. AIIE*, **4**, March, 19 (1972).
61. Davies, B. T., 'Moving Loads Manually', *Appl. Ergonomics*, **3**, December, 190 (1972).

4

Electrical actuation

Practical requirements for actuators

Whatever form of actuator is adopted as the robot muscle, there are certain general requirements. The actuator should have a high ratio of power output to weight. In addition, it should provide the most flexible form of control of movement possible. Any suggestion of jerky movement will, in general, be quite unsatisfactory for the general-purpose robot. Consequently, the output torque should vary smoothly over the whole range of movement. The acceleration of the movement should be as rapid as possible, as should the deceleration. The deceleration can be an extremely important factor when the safety of the robot is considered.

It appears to be better to apply a full operating force, and to control the actual output force by variation of an opposing force, rather than to attempt to control the driving force directly. This has been found to give a smooth action when used with pneumatic controllers.

It is important in robot applications that the weights should all be minimised. These include the weights of the actuator, of the required energy store and of the fluid and conversion equipment used, if a hydraulic control is adopted. At the same time, the efficiencies should be maintained as high as possible.

The actuator should be as quiet in action as possible if it is to be acceptable for a general-purpose robot, and there should be no leakage, of, for example, fluid or battery acid.

Above all, safety must be considered if the robot is to be generally accepted. Consequently, it should be very easy to switch off the actuator and to remove the force which it is applying. It is also

58 Electrical Actuation

desirable that actuators should be lockable when this is required for safety purposes.

Recharging or replacing the energy source should be either automatic or an unskilled task.

In some cases it is an advantage if the actuator can operate on the same power source, or at least on the same form of power source, as does the logical arrangement which controls the actuator. From this point of view, electrical actuators are to be preferred, though sub-units are now available for carrying out logical switching operations on pneumatically operated systems directly without the intervention of electricity.

Whichever of the various possible forms of actuator is adopted, it is always necessary to reach a compromise between the speed of response obtained and the efficiency of the system in any mobile application, since the efficiency determines the size and weight of the power supply required. Selection of a lubricant needs care[25].

In the animal system it appears that there is considerable tremor in units of an actuating muscle but that this is reduced by the summation of the random perturbing tremors. This can be done in the robot. However, it should be remembered that tremor is often beneficial if it is small. For example, on a servo-operated chart recorder it is often beneficial to introduce tremor to reduce the effect stiction between the pen and the paper upon which it writes. The writer has found it useful in some cases to introduce tremor at right angles to the required direction of motion. This reduces stiction without affecting forward accuracy. Rotation of a rod actuator reduces linear movement stiction.

General comparison of storage methods

At the present time it is not possible to state categorically that any one of the possible methods of energy storage for robot use is to be preferred in all circumstances. For example, a major contribution to the total time delay of a system is provided by the time constant formed by the inertia of the load and the stiffness of the system. A comparison of this time constant (t.c.) for different systems[27, 29] is as follows:

System Hydraulic Electric Pneumatic (700 kN/m^2) (2000 kN/m^2)
t.c.   1.0   19.7   31.6   18.25

Now since it is not possible to store energy using incompressible hydraulic fluids, it is necessary to use electrical or pneumatic primary storage. The over-all storage of electrical energy is less heavy than is the conventional storage of pneumatic energy, but the electrical actuator is, in general, heavier and noisier than is the pneumatic actuator.

Klein and Montgomery[28] have compared electrical actuation with pneumatic actuation for a particular task; and they find that while the pneumatic system has a lower initial weight and volume than the electrical system, it needs more refills per day and the daily cost is vastly greater (assumed output, 40 kJ).

Because of such reasons, at the present time for robotic applications where weight and volume, and also noise, are likely to be much less important than is the case with human prosthetic devices, the electrical system is to be preferred.

However, it is rash to generalise, and it is quite possible that future work on prosthetics might change the picture.

Electrical actuation[1, 2]

The use of electric motors for mechanical actuation of robot limbs is very attractive, since a wide range of suitable devices is available, flexible methods of control are known, and perhaps above all electrical power for the recharging of the power storage batteries is almost universally available. Even for use on robot devices in outer space, electrical actuation is desirable, since the electrical storage batteries can be recharged by means of solar-cell photoelectric devices.

Permanent-magnet d.c. motors are an attractive choice for electrical actuation. Not only do they have an efficiency of as much as 60% over-all, as compared with a figure of 40% for a motor having a wound electromagnet field, but they are, in general, cheaper, quieter, smaller and lighter than the wound-field device. Modern motors using ceramic magnets are particularly light.

However, the rotational speed of motors is usually much too high for actuation purposes, and so it is necessary to add a gear-box. Not only does this reduce the speed to a usable value, but it also increases the output force or torque which is available. Unfortunately, the efficiency of gearing is low, varying from about 80% for low ratios to 30% for the higher ratios which are more likely to be necessary in robot applications.

The permissible weight is a most important factor in the selection of an electric motor for robot drive and actuation purposes. The best efficiency of operation will be obtained by operating a motor at about one-half of the maximum power available, and this will minimise the electrical energy requirements and the drain on the power supplies. In addition, the use of a larger motor has the advantage that a high value of peak power is available for use in emergencies or in other special circumstances. However, the additional weight has to be transported continuously. This is a particular disadvantage for some applications—for example, where the robot has to be transported in a space vehicle for outer space applications.

In many applications it will not be permissible to accept the weight (or the price) penalty imposed by the use of larger motors, and the power-to-weight ratio will have to be maximised by the use of a small motor operated at the maximum power output level. This compromise between weight and efficiency is necessary in almost all common applications, and the weight of the required power supply batteries should not be overlooked in such a study.

With some forms of motor the possibility of operation under a continuously stalled condition for considerable lengths of time in robot applications must be accepted. For example, if a robot has to maintain a grip, or to hold up a weight, then the force applied is continuous although there is no movement. Now if a normal shunt type of d.c. motor is stalled on a constant-voltage supply, then the current drain becomes relatively enormous because of the absence of back-e.m.f., which normally limits the current flow. Overload switches can be used to prevent the excessive drain on the supply batteries whenever the supply current exceeds a certain present value; alternatively, mechanical limit switches can be used in some applications in order to break the supply. However, with modern methods of control of motors it is possible to allow operation under constant-current stalled conditions, and there is a great deal to be said for constant-torque, rather than constant-speed, methods of operation. For traction purposes, and also for lifting purposes, constant-torque methods have been widely used. Recently developed methods of constant-torque drive can, however, give a much more flexible method of control than that which has been used in vehicles and in crane and elevator applications[42, 43].

A typical electric motor used in prosthetic applications has been the size 08, 12 V d.c. permanent-magnet motor, which has a stalled torque of 0.012 Nm. This has given a finger-grip of 25 N when driving through gearing and a non-return screw, a 15.6 V 0.9 Ah battery being used for supply.

Where a rotating electrical machine is used as a source of mechanical power in a robot, it is usually necessary to reduce the speed of rotation by a very large ratio. One way of doing this is by means of a worm type of reduction gearing, and this has the additional

Electrical Actuation 61

advantage for many applications that it can be made nonreversible, so that reaction from the load cannot cause the motor to rotate once the supply has been removed; thus further load movement is prevented.

Electrical energy can be conserved by this means, since continuous energising of the drive motor in a stalled condition is not necessary, for example, once the load has been gripped. This feature is most important in a mobile application where the supply must come from electrical storage batteries.

Unfortunately, for some applications the worm gear has undesirable properties. The mechanical losses of a worm gear are high; it is necessary for there to be a change of direction between the input and output shafts; and the loading on individual teeth is high because of the few teeth engaged at any one time. All of these problems appear to have been overcome by the introduction of a simple form of two-stage, in-line epicylic gear, which can give a ratio of, for example, 666:1 with the output and input shafts in-line and both rotating in the same direction.

For some applications the linear form of induction motor provides an attractive form of actuator, provided that a polyphase supply is available or can be generated. The printed circuit motor[3, 20] is a recent development of the early Barlow Disc motor, for d.c. operation. It has a high ratio of torque to inertia, so that the mechanical time constant can be as short as 30 ms, while there is no iron to saturate in the rotor. It appears to be a promising device for robot applications. Low speeds are obtainable without gearing.

Davies[44] has suggested that one of the highest power-to-weight ratios and highest power-to-volume ratios is given by the American 'Globe' type permanent-magnet d.c. motor, continuously rated at 25 W, 12 V. This motor is 80 mm long and 40 mm diameter, and the weight is 0.3 kg.

Although electrical clutches and brakes of many different forms have been developed, the weight and the additional electrical drain required make them undesirable for mobile uses. However, the possibility of using a central electrical or other form of drive motor, with the mechanical distribution of power controlled by means of clutches, might be worth considering in some cases. Certainly there have been numerous developments recently in clutches and brakes—for example, for use in applications to business machines.

It is a strange point that the eddy-current form of brake has received much attention lately, although it is inherently incapable of operation right down to zero speed. Other forms of brake, such as the hysteresis brake, which does not appear to suffer from the same limitation, have received less attention.

The noise of electrical drives

A major problem encountered with electrical drives in some applications is that of noisy operation. In domestic applications the quite astonishing noise levels of some domestic appliances such as washing machines or vacuum cleaners is accepted without question at present, though that of a whirring, noisy mobile robot might be more psychologically undesirable. In many applications, however, the noise level will be quite unimportant.

Where electric motors have been used with prosthetic devices, it has been found possible to limit the noise by using alternate steel and plastic helical-spur gearing with a reasonable precision. The residual noise from the motor is treated by suspending the unit from resilient rubber mountings in order to decrease the transmission of vibrations to the outer structure. In addition, the whole unit is entirely enclosed in a virtually airtight shell, so that airborne noises are retained[15].

On some small motors it has been found possible to reduce the brush and contact noise by covering part of the motor with a muff made from foam plastic. This also has the beneficial effect, if the motor is not totally sealed, of excluding dust. In order to obtain a long life, sintered bronze bearings impregnated with oil are used.

Motors such as these are typically used in a domestic environment for battery-operated record-players, and so they are checked for noise, during manufacture, in a soundproof box containing a microphone.

In applications where the noise level is very important—for example, in robots for hospital use, or for library use—but where the over-all weight of the robot is not of great importance, it is possible to apply the recently developed acoustic cladding. This is formed from a flexible plastic foam about 5 cm thick, coated on the inside with a light resin-bonded Fibreglass skin which is moulded to the shape to be covered. The plastic foam is covered on the outside with an outer skin of lead sheet weighing 25 kg/m^2 which is itself sandwiched between layers of Fibreglass.

In effect, this combination forms a spring-and-mass combination having a low value of resonant frequency, which acts as a filter to the vibration. The cladding has been successfully used on diesel engines, since it can withstand temperatures of 120°C. At a frequency of one kHz a sound reduction of 25–30 dB is obtained.

Electrical Actuation   63

It is important to note that the noise produced by a robot can interfere not only with humans and with sensitive apparatus in the environment but also with the operation of the robot itself. If the robot is fitted with ears, it might well be necessary in some cases to provide the typically animal action of 'freezing' to prevent the sound of the movements of the robot itself from masking external sounds. However, in a noisy environment it is not always easy to separate out the internal and the externally produced sounds, and it would be necessary to take steps to avoid a complete paralysis. This is in fact analogous to the 'cocktail party' problem of hearing. In some environments there is no problem, because the ambient noise is so overpowering that any noise produced by the robot is quite unimportant.

In the latter connection, the writer once spent several weeks at the request of the Admiralty, reducing the noise level of a piece of equipment for installation in the engine-room of a well-known aircraft carrier. It was necessary to work a great deal of overtime, despite the glorious summer weather at the time. The noise level was successfully reduced, but when the equipment was installed it was found that the engine-room noise level was already so outrageous that the slight noise of the equipment would not have been heard anyway. In this case the tax-payer had to find the cost of the extra work.

In robot applications it is desirable to keep a sense of perspective about noise levels and to consider carefully whether or not humans are likely to be inconvenienced by the noise. If there are no humans in the operational environment, then the only reason for noise reduction is that it can often indicate the presence of excessive mechanical wear. However, noise reduction is likely to be an important factor in the acceptance of the robot for domestic use[15-17].

Solenoids as actuators[4-9]

The solenoid is a useful device where linear motion is required in robot applications, and various forms and power ratings are now freely available. The characteristic curve of force plotted against distance moved depends on the length and shape of the iron plunger which is pulled magnetically into the solenoid. If the force is required to be reasonably constant with the distance moved, then the plunger should be long. A short plunger will give a sharper peak in the force–distance characteristic, assuming that the current energising the solenoid is maintained constant.

The use of a frame made of magnetic iron surrounding the coil has

64 Electrical Actuation

little effect on the pull exerted at the beginning of the stroke, but it causes a large increase of the force at the end of the stroke.

The maximum force exerted by a solenoid is given by

$$\text{maximum force} = \text{constant} \times \frac{\text{C.S.A.} \times \text{turns} \times \text{current}}{\text{length}}$$

where C.S.A. is the cross-sectional area. The electrical energy required for a solenoid is at least twice the mechanical output energy obtained.

The characteristics of short-stroke solenoids are sometimes improved by shaping the end of the iron armature. For example, a popular shape is an armature with a conical end which moves into a hollow conical cavity in the fixed part of the iron circuit.

If the pull is required to be of high value when the solenoid is closed, it is an advantage to reduce the size of the mating iron surfaces —for example, by making one of them dome-shaped. This method is based on the fact that if the value of the flux is fixed, then the pull is proportional to the square of the flux divided by the cross-sectional area. In some cases mechanical toggles are used to obtain a high value of closed (or sealed) pull.

The coil of a solenoid is inductive and the inductance changes as the solenoid core moves on actuation. Consequently, if a constant voltage is applied to the coil of a solenoid, a fairly complex waveshape of current is obtained.

Initially the current builds up exponentially because of the inductance of the open solenoid. However, once the core starts to move a back-e.m.f. is produced which opposes the current flow and so causes the current value to fall. The movement accelerates; but when the movement is finished and the solenoid has moved to the limit, then the current again continues to build up exponentially. Since the inductance is now greater, the current builds up towards its final value more slowly than when it was open. Care is required when inductive loads such as solenoids are switched by the use of semiconductor devices.

It is useful to fit some robot devices with magnetic or electromagnetic hands where iron loads must be handled. In some cases these can be used to cause the separation of, for example, steel sheets by mutual repulsion.

In most robot applications solenoids will be actuated by direct current. If, however, a.c. actuation is required, then the magnetic circuit must, in general, be laminated. In order to avoid vibration and buzzing, it is usually necessary to fit shading poles. These involve the enclosure of part of the pole face with a closed electrical circuit, formed, for example, from a ring of copper. This is, unfortunately, a point of unreliability in practice, for it is not unknown for the shading pole rings to break and so allow vibration and noise to appear. An additional factor in a.c. energised devices is that the pole faces must be very carefully ground flat if vibration is to be avoided[36].

The writer has discussed elsewhere some of the problems encountered when inductive loads such as solenoid valves must be switched by transistors[37, 38], by cold cathode tubes[39], or by magnetic amplifiers[40, 41].

Torque motors and rotary solenoids

Where a limited angular rotation is required to be produced electrically in robot applications, it is sometimes possible to use the rotary solenoid. Rotary solenoids of the Ledex type have been widely adopted[10], particularly for the actuation of rotary switches. In this type of solenoid a coil in an iron enclosure provides an axial magnetic pull to an end plate. Steel balls resting between the end plate and inclined tracks are used to convert the axial pull of the electromagnet into a rotary motion of the end plate. As an example of the application of these rotary solenoids, the writer has incorporated them into numerical control equipment for machine tools.

More recently developed devices have incorporated true rotary solenoids in which iron rotor teeth are pulled circumferentially into the gap between inner and outer stator teeth when the coil of an electromagnet is energised. This device has the advantage of possessing an armature which is both light in weight and balanced, these factors reducing the effects of shock, acceleration and vibration. In addition, there is little radial or axial load on the bearings, which can be both substantial and widely spaced. The arrangement adopted gives a constant torque characteristic, though it is possible to construct devices which give an inverse square law form of torque characteristic.

The advantage of such solenoid devices for use where a limited rotary motion of, say, 35° is required for robot actuators lies in their compact and completely sealed nature, and they can be made quite easily replaceable for maintenance purposes if required. The disadvantage for mobile use is the very high power requirement of, say, 15 W for a torque of less than 0.1Nm. Continuous stepped rotation is possible if a ratchet mechanism is added. If such a ratchet-and-escapement is fitted then continuous energisation is unnecessary, so that for some applications the average power dissipation requirement can be quite small.

For use in servo applications in which a limited amount of angular motion is required the rotary torque motor has been developed. This has a torque which is a linear function of the exciting current. As an example, a typical device has a permanent-magnet rotor and an epoxy-encapsulated stator with an aluminium outer ring. Typical relevant data for such a device are as follows:

Weight	0.1 kg
Continuous rated torque	0.01 Nm
Angular motion: torque falls to	92% at ±60° rotation ⎫ at
	80% at ±70° rotation ⎬ max.
	45% at ±80° rotation ⎭ current
Power input for 0.03 Nm torque	8 W at 25°C
Resistance	48 Ω at 25°C
Sensitivity	0.1 Nm/A
Electrical time constant (L/R)	8×10^{-4} s
Back-e.m.f.	0.07 V rad^{-1}s^{-1}
Maximum winding temperature	155°C
Approximate size	0.05 m dia × 0.02 m long

It should perhaps be mentioned that the Ledex or the uniselector type of movement is inherently only capable of providing rotation in one direction. If reversible rotation is required, two solenoids are normally used, with gearing on each which engages automatically when the solenoid moves. It is possible to use the differential movement of two solenoids, and this has been done with uniselector devices, though usually to achieve an increase and decrease of a switched resistance value rather than a movement. An alternative is the adoption of a reversing gear-box, though this is likely to be somewhat heavy, since it must be electrically actuated.

It is advantageous in a rotary stepping device to adopt the uniselector principle of stepping on the release of the electromagnet rather than on the energisation. A high force is then available right at the commencement of the operative stroke, being provided by a spring which is charged by the energisation of the magnet. This principle is now adopted in other stepping motors[11].

The Steromotor is basically an alternating current motor in which the rotor is not on conventional bearings. Instead it rolls around a track inside the stator. Because of this rolling action, the effective output speed is low, since a flexible linkage is used to translate the rolling motion of the rotor into a rotary motion of the output shaft. Basically, this is an a.c. motor of the induction type, with fixed stator coils and a permanent-magnet rotor. For use as a stepping motor, pulses of direct current can be applied, up to 1200 steps per revolution being possible.

The Steromotor has been found useful in a.c.-operated actuator schemes for several reasons. Although the rotor inertia as seen at the output shaft is negligible, there is nevertheless an inherent braking torque whenever the supply voltage is removed. With d.c. energisation there is a high braking torque. With a.c. drive it is sometimes a useful feature of this motor that the input impedance has an almost constant value at all speeds from the full rated speed down to the stalled condition. Operation indefinitely in the stalled condition is quite safe. Acceleration and braking are very rapid, time constants of the order of 0.02 s being achieved. Such motors show promise for application to robot device actuators.

Step-servo motors[12-14, 18, 19, 26, 32, 34, 35]

Because of their importance in industrial applications, step-servo motors are receiving an increasing amount of attention. For example, the writer's students are becoming very interested in the application of the step-servo motor to the numerical control of machine tools and industrial processes.

There are two basic forms of step-servo motor. The permanent-magnet form uses the reaction between an electromagnetically produced field and a permanent magnet. The variable-reluctance form uses the solenoid principle of reaction between an electromagnetically produced field and a piece of soft iron. In both forms a number of solenoid coils are arranged in a circle, and sequential energisation of these coils causes the rotor to move round in a series of steps.

The advantages claimed for such devices over other sources of motive power include the independence of the amplitude variations of the applied voltages or pulses, the insensitivity to shock and vibration and the long life of up to one thousand million operations. In addition, the response of the device is fast, the response time being as low as 1 ms. The amount of motion per step can be maintained very accurately constant by the use of magnetic detent devices.

There are various sequences in which the windings of a step-servo motor can be energised in order to cause it to rotate. One way in which the various possible sequences have been compared is the use of the ratio of the ampere-turns of energisation to the necessary supply power dissipated in the windings, as a figure of merit. A reduction of the required power for a given excitation not only has the great advantage for mobile applications such as robots that a lower weight of supply battery is required; it has the additional advantage that the step-servo device runs cooler, and the reduced temperature helps to increase the life of the device. However, some possible sequences require very complex forms of switching sequence control arrangements, and this is undesirable from the points of view of cost, complexity and reliability.

The determination of the type of step-servo to use for any particular application can be based on considerations such as the following. The permanent-magnet device gives magnetic detenting and large stepping angles and it is suitable for use where non-ambiguity is required and the pulse-rate is below, say, 300 pulses/s. On the other hand, the variable-reluctance device is suitable for use when magnetic detenting is not required but small stepping angles are needed without the introduction of gearing, and the pulse rate must be as high as 1200 pulses/s. It is also suitable where some ambiguity of the position is unimportant, and particularly where, because of the presence of iron dust or particles, permanent magnets are not permissible.

Cooling[21-24, 30, 31, 46]

It is an advantage with a general-purpose robot to have the over-all size as small as possible. For this reason, the adoption of modern techniques of miniaturisation is essential. There are two main difficulties encountered when electrical equipment is miniaturised. The first is that of the achievement of a good degree of electrical insulation, and the most important point here is the uniformity of the insulating material used.

An even more important difficulty at the present time is that of removal of the heat generated by any electrical equipment and the maintenance of a reasonably low temperature of operation under all conditions. It is important to minimise the power lost because of heating, since this power has to be provided from the primary power source, which must be carried by a mobile robot. In addition, if the temperatures of some parts of the robot are allowed to become excessive, then the properties of the materials from which the robot is made can change drastically and so introduce a source of unreliability.

If such excessive heating is to be avoided, then the primary need is to keep to a minimum the number of watts of heat energy dissipated per unit volume of the equipment, and this involves the use of techniques and equipment having the highest possible electrical efficiency. Although there is a continuous increase of effective efficiency as equipment is improved, there is at any one time a limit to the value which can be achieved. Thus at any given time there exists a given definite minimum to the quantity of heat which must be produced by electrical losses in the equipment, while at the same time there is a given definite maximum to the internal temperatures which can be tolerated. If the robot is to be reliable, then these two opposing factors must be reconciled.

One way of reconciling them is by the use of a large area of cooling surface exposed to the ambient air or fluid. In outer space it is necessary instead to provide a radiating surface. Between the device producing the electrical losses causing the heating and the main cooling surface (or heat-sink) there must be a path having the lowest possible thermal resistance to heat flow. It is important to note that for reliability it is not only the mean temperature of apparatus which must be limited, since failure is likely to occur at any 'hot spot' which appears. Other problems can be caused by the cooling means adopted, such as a reduction of the electrical breakdown strength, or an increase of the stray capacitance level, and all such problems must be considered carefully at the design stage.

Wherever possible, cooling by means of the ambient air is desirable, and a wide range of extruded forms of alloy heat-sinks is now available for this purpose. If it is necessary to isolate the equipment electrically from the heat-sink, then it is usual to use insulating screws and thin mica washers to minimise the temperature drop across the insulation. Well-designed liquid cooling systems have advantages, but they can be undesirable in mobile applications. This applies also to 'boiling' types of cooling system, which make use of the latent heat of evaporation of a liquid.

A logical extension of evaporative cooling is the adoption of some form of perspiration cooling. However, such a method would seem to be undesirable in the general-purpose robot from many points of view. The method could not be used in some forms of environment, such as dusty or highly humid atmospheres. And the prospect of a domestic robot which suffered from B.O. is obnoxious in the extreme!

With a pneumatic system the working fluid must be stored and carried in bottles at a high pressure in liquid form. Great care must be taken to ensure that the temperature of the bottle cannot rise too high, since this would cause a dangerous increase of pressure. An additional problem is that any leakage of the gas from the system will cause a local fall of temperature, and it is possible for the stored liquid to freeze (typically at $-55°C$). A human touching the bottle would then be severely burned. An escaping jet of gas can likewise cause a burn. At the same time, some form of relief should be provided to protect against explosion in the event of the pressure increase mentioned above.

In some cases—for example, when a robot has to operate on the

70 Electrical Actuation

cool side of the moon—it will be necessary to provide means for heating rather than for cooling. It is clearly desirable that a dual-purpose temperature stabilising system, which can either heat or cool, should be fitted in such cases of extreme temperature variation. Motors and fans used in lunar equipment have been cooled with the aid of heat-dissipating coatings of beryllium oxide microspheres in silicone rubber[33].

REFERENCES

1. Wilson, R. R., 'Electric Motor Operated Actuators and Their Control', *Laurence Scott Eng. Bull.*, **9**, 17 (1967).
2. Mitchell, W. S. E., 'U.K. Developments in Myo-Electric Powered Prostheses', *Des. Electron.*, **5**, July, 16 (1968).
3. Rothwell, D., 'The Printed Circuit Servo Motor', *Prod. Des. Eng.*, August, (1964).
4. Taylor, D., 'Electric Actuators', *Control & Instrum.*, **2**, June, 25 (1970).
5. Say, M. G., (ed.), *Electrical Engineer's Reference Book*, 225, Newnes (1968).
6. Fink, D. G., *Standard Handbook for Electrical Engineers*, 5.4, McGraw-Hill (1968).
7. Walker, R. C., *Relays*, 38, Chapman and Hall (1953).
8. Anon., 'British Solenoids for Control Purposes' (Review), *Automn Prog.*, April, 114 (1960).
9. Burdett, G. A. T., *Automatic Control Handbook*, 2.1, Newnes (1962).
10. Young, John F. and Wootton, R., 'Progress on a Numerical Control System', *G.E.C. Internal Rep.*, SWD 72 (1959).
11. Cullen, G. W., 'The Design and Development of the Rotenoid', *Component Technol.*, **2**, March, 20 (1966).
12. Victor, R. A., 'Memory Logic Allows Stepping Motor to Catch Up', *Control Eng.*, **8**, February, 143 (1961).
13. Kieburtz, R. B., 'The Step Motor—The Next Advance in Control Systems', *Trans. IEEE*, **AC9**, January, 98 (1964).
14. Delgado, M. A., 'Mathematical Model of a Stepping Motor Operating as a Fine Positioner Around a Given Step', *Proc. Joint Autom. Control Conf. A.A.C.C., Colorado, 1969*.
15. Harris, C. M., *Handbook of Noise Control*, McGraw-Hill (1957).
16. Beranek, L. L., *Noise Reduction*, McGraw-Hill (1960).
17. Allen, C. H., 'Guidelines for Designing Quieter Equipment', *Mech. Eng.*, **92**, January, 29 (1970).
18. Bell, R., et al., 'The Use of Stepping Motors in Numerically Controlled Machine Tools—A Summary of the Present State of Development', *Int. J. Mach. Tool Des. Res.*, **10**, December, 417 (1970).
19. Fredriksen, T. R., 'The Closed Loop Step Motor', *Automatica*, **5**, 61 (1969).
20. Berg, M., 'Printed Circuit D.C. Motors', *Elektron. Rdsch.*, **17**, November, 582 (1963).
21. Johnson, D. P. and Silcock, R., 'Cooling for Reliability', *Proc. Reliability in Electronics Conf., IEE, December, 1969*, 114.
22. Shaw, E. N., 'Heat Control in Electronic Equipment', *Electron. Eng.*, **29**, January/February/March, 13, 65, 115 (1957).
23. Shaw, E. N., 'Liquid Cooling of Electronic Equipment', *Electron. Eng.*, **30**, September, 516 (1958).
24. Arcus, A. A., et al., 'Testing for Chemical Inertness in Electronic Coolants', *Wescon Conv. Rec.*, **13**, paper 2/4 (1969).

Electrical Actuation 71

25. Vest, C. E., 'Development and Use of an In Situ MoS$_2$ Solid Film Lubricant', *Wescon Conv. Rec.*, **13**, paper 3/1 (1969).
26. Head, J. R., 'A Versatile Stepper-Motor Drive Unit', *Des. Electron.*, **7**, June, 63 (1970).
27. Lambert, T. H. and Hall, M. J., 'Design and Control of Powered Artificial Arms', *Proc. Instn Mech. Engrs*, **183**, pt 3J, November, 1 (1968).
28. Klein, P. M. V. and Montgomery, S. R., 'Portable Power Supplies for Prostheses', *Proc. Instn Mech. Engrs*, **183**, pt 3J, November, 26 (1968).
29. Davies, R. M. and Lambert, T. H., 'Dynamic Characteristics of Pneumatic, Electric and Hydraulic Actuation of Prosthetic and Orthotic Devices', *Proc. Int. Symp. External Control Hum. Extremities, 1966*, 65.
30. Cook, D. V., 'Cooling Electronics Naturally', *Mach. Des.*, **42**, August 20, 126 (1970).
31. Schenck, H., *Heat Transfer Engineering*, Prentice-Hall (1959).
32. Chiarella, L. J., 'Rotation by the Digits', *Machine. Des.*, **42**, November 26, 84 (1970).
33. Anon., 'Heat Dissipating Insulating Encapsulant Used in Lunar Landing Equipment', *Insulation/Circuits*, **17**, May, 52 (1971).
34. King, D. S., 'Stepper Motors for Digital Control Systems', *Control & Instrum.*, **3**, June, 29 (1971).
35. Olson, W. L., 'Stepper Motor Replaces a Servo Motor', *Electromech. Des.*, **9**, October, 44 (1965).
36. Young, John F., 'Methods of Semiconductor Static Switching', *Electl Rev.*, **174**, April 17, 591 (1964).
37. Young, John F., 'The Switching of Inductive Loads Using Transistors', *Br. Comm. Electron.*, **10**, November, 844 (1963).
38. Young, John F., 'The Use of Transistors in Industrial Timer Circuits', *Electron Eng.*, **35**, June, 366 (1963).
39. Young, John F., 'Comparison of Cold-Cathode Tubes with Transistors', *Br. Comm. Electron.*, **11**, July, 484 (1964).
40. Young, John F., 'Ferro-Resonance, Problems and Applications', *Electl Rev.*, **176**, May 2, 782 (1965).
41. Young, John F., 'Using Capacitors to Improve Low-Cost Magnetic Amplifiers', *Engineer, Lond.*, **220**, July 30, 176 (1965).
42. Young, John F., 'Field Control of Motors with Constant-Current Armature Supply', *Control*, **12**, January, 35 (1968).
43. Young, John F., 'Control Possibilities of Double-Stator Squirrel-Cage Motors', *Control*, **12**, May, 416 (1968).
44. Davies, B. L., 'A Prototype Portable Hydraulic Power Supply for Prosthetic Applications', *Proc. Conf. Hum. Locomotor Eng., Sussex, September, 1971* (Instn Mech. Engrs), 285.
45. Alderson, S. W., 'The Electric Arm', in: Klopsteg, P. E. and Wilson, P. D. (eds.), *Human Limbs and Their Substitutes*, 359 (see 403), McGraw-Hill (1954).
46. Breese, J. C., 'Check before You Freeze that Design', *Electron. Des.*, **21**, January 4, 90 (1973).

5

Power for actuation

Battery economy[1]

In a normal servo system, if a steady-state error exists, then the output motive devices must be continuously energised in order to oppose the error. Now in a robot limb or hand a steady-state position 'error' is quite usual—for example, when a hand is gripping some object with a certain force. It is undesirable that there should be a drain on the power supply source in such a case. In the case of pneumatically or hydraulically operated devices there is no steady drain in such cases provided that the pressure is maintained. However, in the case of an electrically operated hand, there is a continuous drain on the supply batteries if the force is to be maintained by the electric motor which drives the hand.

This is most undesirable in the case of a mobile robot, and so some form of non-reversible transmission such as a worm-gear drive should be used. The drive motor can then be de-energised as soon as the correct force has been achieved, as determined by the input to the servo system. Any variation of the input then causes the drive motor to be re-energised and the worm gear to be driven again, in either the forward or the reverse direction.

There are other advantages in the use of such a worm drive and de-energisation arrangement. The servo system can be of the simple 'bang-bang' or 'on–off' type, so reducing the power requirements and losses in the amplifier controlling the motor and incidentally making possible an improvement in the rise-time of the output actuator force. In addition, the system is unconditionally stable in this 'gripped' condition, since the drive motor is then disconnected from the moving parts in effect. A compromise is necessary between the effective width of the dead zone introduced by the action described and the gain of the servo loop. It is desirable that the extent of the dead zone should be small, but if it is too small then an excessively high loop gain is required to ensure that maximum drive torque is available at the limit of the zone. The high loop gain makes the system performance less dependent on variations of the gain, of the load and of the static friction of the moving parts, while at the same time the response time of the system is reduced. Unfortunately, a high value of loop gain also increases the problem of avoidance of servo instability and the resulting hunting, which would be disastrous in a robot mechanism if it was allowed to build up.

In prosthetic devices incorporating such a dead zone for battery economy, suitable values of dead-zone input limit have been found to be about 30 mV input, with a nominal gain in the amplifier of about × 500. Such quenching circuits in prosthetics have been designed to continue to act even though the supply battery is almost exhausted, and to be capable of applying a grip of 30 N under this condition. To prevent a dead-zone form of instability, it has been found desirable to introduce an additional 50% hysteresis effect as shown in *Figure 3.6* (p. 53).

In the well-known Russian powered artificial hand the fingers can be opened or closed at maximum speed, and there is no feedback other than visual feedback to the human operator. However, effective speed variations can be obtained by 'twitching' the motor, the resulting impulsive speed variations then being smoothed out by the mechanical inertia of the system. The quality of control is then rather poor. In addition, the electrical consumption from battery is very high during movement, though it is low in the quiescent state.

Clutches and brakes

Many different forms of clutch[11, 16] and of brake have been used. In a robot the clutch or the brake must be capable of operation from the control system by means of the electrical or other signals which are provided.

In general, electrical forms will be used. The forms which use electromagnetic forces to produce mechanical movement can sometimes be used, but more recently developed devices are sometimes preferable. One form incorporates magnetic powder (or fluid) which can, in effect, be solidified electromagnetically. Another form is the well-known eddy-current brake. However, this is a most undesirable form, since it is inherently incapable of operation down to zero speed. The hysteresis form would appear to be preferable from this point of view.

74 Power for Actuation

In general, in robot applications it is desirable to use the approach of positive excitation and inhibition and, consequently, to achieve braking by the application of a reverse force. However, in some cases this will be ruled out because of the space and weight of the complete equipment, and in such cases clutches and brakes might be fitted to robots.

It will sometimes be necessary to avoid the use of a dynamic speed control, and in such cases the simple flying-weight form of mechanical governor will be applicable or possibly, for greater reliability, the permanent-magnet eddy-current form. With modern magnetic materials this need not be excessively heavy.

It is always necessary to dissipate power in the form of heat in clutches and brakes. This is most undesirable in the mobile robot, since this power must be supplied from the prime power source which must be carried by the robot. Consequently, any power dissipated as heat has the effect ultimately of increasing the weight and the size of the mobile robot, as well as decreasing the efficiency. It is better to use switched or regenerative methods of speed control wherever this is possible.

The heat produced in clutches and brakes, where these must be used, has to be dissipated in some way. There are in theory possibilities of using this heat to regenerate electricity, but this course is not practicable. Where the robot operates in an environment of air at ambient temperatures which are not excessive it is possible to design the magnet system to enable simple natural air cooling by encouraging air to flow behind the friction surfaces and across the coil. This is not possible for robots which must operate in other environments, however.

The magnetic powder clutch is suitable for use at very high power level, but it has the disadvantages that the powder is abrasive and is not uniformly distributed. The magnetic fluid clutch is suitable for use up to 300 W and it gives a power gain of up to 200:1. Not only does the fluid cool and lubricate the clutch, but it ensures smooth transition. However, there is a sealing problem, which is unfortunate in domestic use.

In general, it is best where possible to avoid the use of both clutches and brakes in mobile robots.

Pneumatic actuation[20, 21, 23, 26]

With an open-loop type of control system the low stiffness of the usual pneumatic control devices can introduce problems. It is not easy to provide small corrections necessitated by, for example, load disturbances. The use of a closed-loop form of control system can improve the control and increase the effective stiffness, though there are still problems introduced by the low gas pressures commonly used and by the possibility of instability of the closed-loop control system.

Carbon dioxide, which might possibly be available in very lightweight plastics or fibre containers, is probably the best form of supply source. A pressure reducing valve is used in the supply system, so that the carbon dioxide pressure is reduced from the high

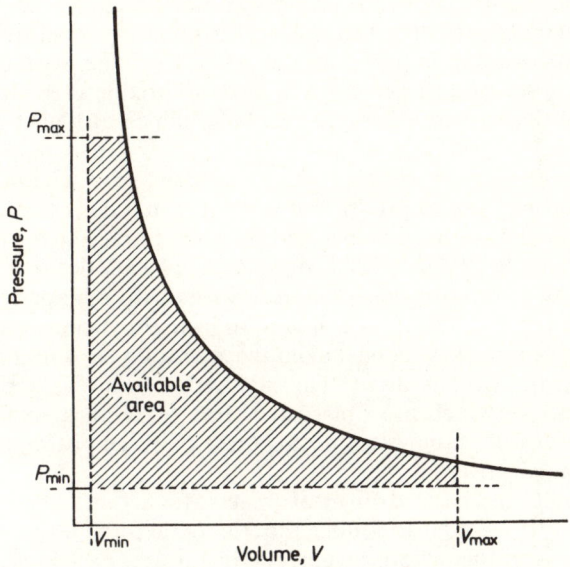

Figure 5.1   Pressure/volume pneumatic actuation curve

value of 5250 kN/m^2 down to about 600 kN/m^2. If the pressure is plotted against the volume for isothermal conditions, a curve of the form of *Figure 5.1* is obtained. The area under such a curve represents the theoretical amount of work which can be obtained for a fixed mass of gas. Now if the pressure cannot exceed the level of the dotted line, all of the work represented by the area above the dotted line is lost. In addition, there is a maximum volume, and so all of the work represented by the area to the right of the vertical dotted line is lost also. Very low pressures cannot be used; nor can very low values of volume where compressibility is lost. Consequently, only the central area can be used, and in practice this area represents only about

24% of the available energy. In addition, at reduced load the efficiency falls, since the same quantity is required as at full load. There is a possibility of improving the position by the use of the more complex differential actuators in pneumatic systems.

With a piston-operated system the friction of the piston in the cylinder introduces some differential between the pressures required for movement in opposite directions. This differential is largely independent of the load, though there may be some variation at high values of load, probably because of deformation of the piston sealing rings, which are usually made of rubber. In addition, the friction increases with the usage of the actuator, but eventually settles down to a steady value. Molybdenum-disulphide-based greases are used.

Experimental results with such pneumatically operated devices show that typically, at a pressure of 500 kN/m^2, the work obtainable is of the order of 20 J per gram of carbon dioxide used. It has been suggested that improved results can be obtained at higher pressures, of the order of 1500 kN/m^2.

It is necessary to restrain the movement of a piston-operated device in some way to prevent movement from one extreme position to the other. A spring can be used for this, though it is more usual to use a single piston with a double-ended cylinder and two-way seals, or two opposing pistons, each in its own cylinder. It is very convenient to use this approach, and to have the two pistons coupled by means of a cable which is taken twice around an actuating pulley, providing the output drive. This method avoids the variations of output torque which are obtained if levers are used to obtain the output. Since the cable is connected directly to the piston, there is no additional rod friction.

A recently introduced form of pneumatic actuator uses a rolling seal principle in order to reduce the friction and the leakage between the piston and the cylinder wall. The piston is a very loose fit in the cylinder, but leakage of gas past the piston is prevented by a flexible Neoprene seal. This seal is folded back upon itself so that it rolls. In effect, when gas is introduced into the cylinder it blows up a Neoprene balloon which is folded over the sides of the loose-fitting piston, and in so doing it is able to force the piston along. Because of the loose fit of the piston and the sealing of the flexible folded-back diaphragm, friction is reduced while leakage past the piston is eliminated. Piston rings are unnecessary, and so is any lubrication. Leakage can only occur along the shaft or actuating rod, and even this is only possible on one side of a double-acting cylinder and not at all with a spring-return single-acting cylinder. Operation is possible with pressures up to about 900 kN/m^2 and over a temperature range of $-40°C$ to $+225°C$. The absence of lubrication is an attractive feature for robot applications. Clearly, this rolling-diaphragm principle can be adapted for use with other gases or fluids.

In the analysis of pneumatic control systems—for example, for prostheses—it has been usual to make various simplifying assumptions such as viscous friction, all-adiabatic processes, perfect gas and complete absence of leakage[23]. The feedback stabilisation of pneumatic prosthetic control systems, using mechanically signalled valves, has been considered by Lord and Chitty[31].

Experience with prosthetics

The McKibben muscle is an ingenious pneumatic device. It takes the form of a rubber tube covered by a criss-cross nylon braid. When the tube is filled with gas, the action of the braid causes the muscle to shorten, the over-all length reduction being about one-third. This muscle is not without problems, however. For example, there is some bounce in the action of the muscle; there is not much power available; and the power decreases as the muscle shortens. Nevertheless, it may be that developments of this device making use of a more cellular structure will become of importance in robot applications.

It should be noted that even if pneumatic actuation is used, it will probably be necessary to include electrical supplies for the control system. There is now, however, some possibility of building completely pneumatic control systems by making use of the newer fluidic control and logic devices.

In general, as Donald[39] has pointed out, pneumatic devices have the advantages of a high power-to-weight ratio, of small control valves and of the availability of high power (though at the expense of gas economy). In comparison, for example, electrical actuation has the disadvantage of a low power-to-weight ratio, of noisy operation due to gearing and of the great weight of the motor and the required gear-box.

Relative disadvantages of pneumatic operation include the need for two separate power sources, electrical and pneumatic as mentioned above; in addition, for some applications the compliance of the gas is a disadvantage.

The feedback of the transient pressure has been found to be a useful method of preventing oscillatory response in some prostheses[15].

The widespread recent availability of various forms of aerosol spray operated by Freon-12 gas might possibly make this a valuable source of power, in expendable containers, for future use in actuators.

One problem with the use of gas stored in liquid form is that the energy required for evaporation must come largely from the liquid,

78 Power for Actuation

and this can cause freezing if the mean power level is allowed to become too high.

It is of interest to note that the energy density of butane gas is extremely high, the specific capacity being some 14 500 Wh/kg, compared with less than 500 Wh/kg for the best electrical storage means. The major difficulty with the use of this power source is its conversion into a usable form. Future work might possibly make it usable, though it is difficult to envisage its use in space. A further future possibility is the use of nuclear power.

Various forms of pneumatic valve have been used in the control of prosthetic devices, such as the Kiessling valve and the Hendon valve. The latter has been shown not only to be very light but also to give a very linear characteristic of flow rate against displacement[19].

It has been found advantageous to convey pneumatic power using nylon tubing with an outside diameter of 3 mm and an inside diameter of 2 mm, with a burst pressure of 12 MN/m^2, though all bends must be of at least 6 cm radius[20].

Pneumatic actuation at Aston

Typical of work on limb control is the investigation carried out at Aston by Donald[39]. The factors considered in this work include:

1. Power economy.
2. Availability of output power.
3. Minimisation of cost and of weight.
4. Accuracy.

Since the factors to be considered are likely to be to some extent incompatible, it is necessary to attempt to achieve an optimisation by balancing the factors against one another.

From the control viewpoint, any device driven by a compressible gas and having a pivoted linkage is inevitably nonlinear in action because of the compressibility of the gas, of saturation and of the existence of dead zones and hysteresis in the mechanical system. Donald used the describing function approach to design, and he incorporated force feedback from the output using strain gauges.

Pulse width modulation of the pulse supply to the control solenoid was found to be the best method of control, the methods of pulse amplitude modulation and pulse frequency modulation not being adopted mainly because of the limited available supply voltage value. A pulse frequency of 10 pulses/s was used.

The over-all closed-loop force control system was as shown in

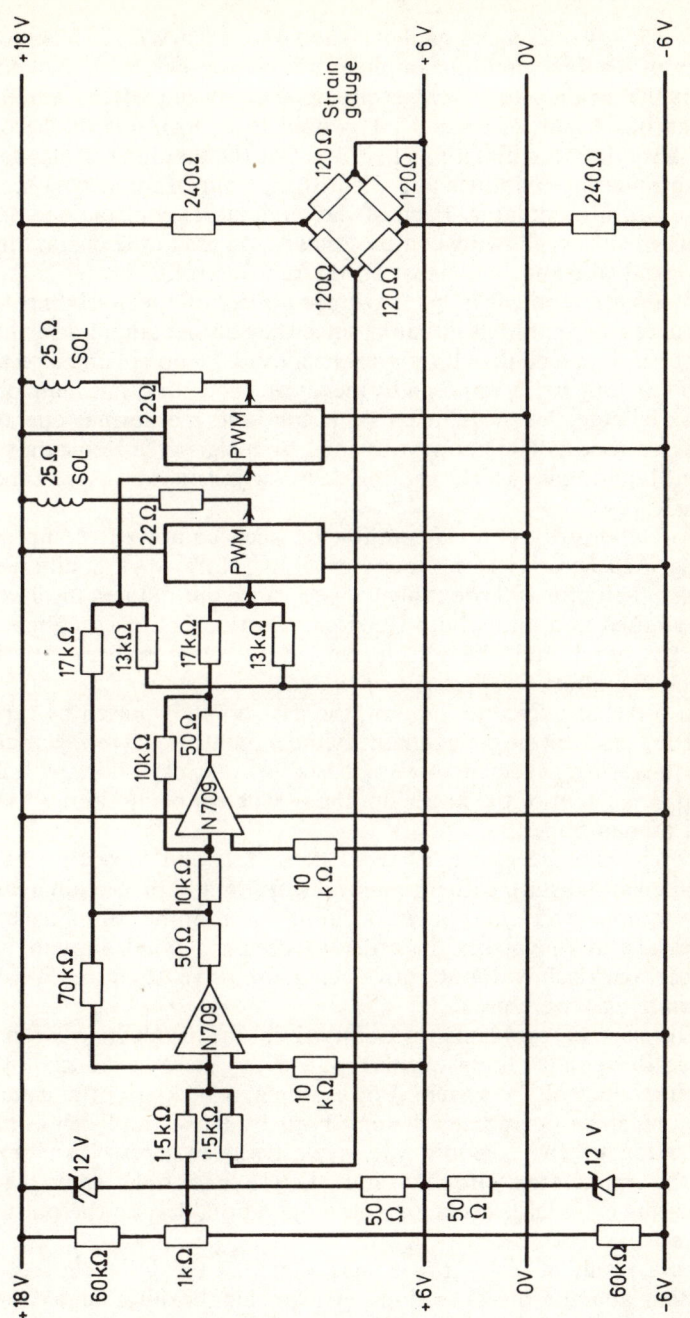

Figure 5.2 The Aston limb control circuit produced by Donald

Figure 5.2. A reference signal of -5 to 0 to $+5$ mV is supplied to the input of the first summing amplifier, for comparison with the voltage from the strain gauge bridge circuit. The output of the amplifier, which has a gain of about 700, is used to control one of the pulse width modulators directly. The output of the amplifier is also taken to an inverting amplifier and then to the control input of a second pulse width modulator. Each of the modulators controls one of the solenoid valves, allowing compressed gas to pass to the piston for the driving stroke and to exhaust on the return stroke.

Pneumatic control is by two-stage valves of the Kiessling type. These can be mounted in banks since they are rectangular in shape. For operation a control lever must be moved 0.6 mm in any direction, a pull of 130g being imposed by means of a solenoid. The main pneumatic cylinder has a diameter of 11 mm and the normal operating pressure used is 600 kN/m² obtained by means of a reducing valve from liquid gas which is stored in cylinders at a pressure of 7000 kN/m².

For laboratory experimentation on such equipment compressed air is used because of its lower cost. In future work a differential form of actuator will probably be used, since this reduces the amount of gas used and limits both the nonlinearity effect of compliance in the gas used and the effects caused by temperature variations which can affect the pressure of the gas.

In order to overcome stiction, the gas supply is pulsed by rapidly opening and closing the main cylinder control valve electrically. The finger arrangement used is based on that of Rakic, with 250° total movement of the finger tip, the movement being similar to that of a human finger.

Force transducers are fitted in the middle and fingertip sections (median and distal), with mechanical stops to give protection against, for example, accidental blows or falling over. Some compliance was necessary in the pivots, in order to obtain a good enough fit to reduce backlash without introducing excessive stiction. Foil-type strain gauges were used.

The sort of problems encountered in the production of pneumatically actuated limbs of this type are, in general, produced by the nonlinearities of the system. For example, a backlash of as much as 15% must be overcome; there is both hysteresis and delay in the operation of the solenoid valve; electrical hysteresis can appear in the modulator circuits; and stiction and tight spots in the actuating cylinder prevent a smooth operation, despite the pulsating drive.

As a result of tests, it was suggested that the joint between the median and the distal sections of a pneumatically actuated finger could be replaced by a permanent bend of 20°, with very little loss of function and considerable simplification to a digit mechanism, as well as the reduction of friction and backlash. In addition, the simplification would ease the mounting of the strain gauges.

Such work as this has done no more than lay the foundations for future investigations, and much work remains to be done on efficient and useful limbs and their actuators.

A form of pneumatic valve for prosthetic use has been described by Cool and Pistecky[28]. This makes use of the expansion of a hot wire when electrically heated, and it is very small (3 cm by 0.3 cm) and very light (1.5 g). Step response times of between 30 and 150 ms have been obtained, some of this time being required to reduce the initial tension in the wire. The flow is proportional to current between 75 and 100 mA, the control power required being less than 0.2 W. This valve has been used in the control of a prosthetic hand and a prosthetic elbow.

Hydraulic actuaiton[13, 20, 21, 40, 41]

The use of hydraulic actuators in robots is worth considering because of several potential advantages. Because of the low compressibility of hydraulic fluids, the stiffness of a hydraulic actuator is high, and this leads to an associated high natural frequency and a rapid response. The ratio of power available to the weight of the system is high for a hydraulic system; the noise level is low; and hydraulic techniques are well-known and safe in operation. Reliable and simple hydraulic actuators are easily available.

However, there are a number of problems with hydraulic systems of actuation. The nature of the hydraulic fluid, high density and low compressibility, make it an unsuitable medium for the storage of energy. Consequently, it is necessary to store the energy required by hydraulic actuators in some other form—for example, pneumatic, electrical or chemical. Leakage of hydraulic fluid from an actuator is most undesirable[27]. In an industrial environment leakage captures dirt and dust and so necessitates frequent cleansing. In a domestic environment hydraulic fluid leakage is even more undesirable. Even leakage inside the system—for example, past valves or past piston rings—is most undesirable, since like external leakage it can cause a reduction of the efficiency of actuator control.

In the general hydraulic control system it is not easy to obtain a high efficiency at the same time as large actuator forces, and a compromise is necessary. It is a possibility that the conduction and induction forms of pump[3] which have been used for liquid metals

82 Power for Actuation

may in the future be developed to the point where they can provide a useful source of hydraulic power for robot use.

One successful application of hydraulic actuation has been the four-legged walking machine[2] developed by the U.S. Army and by General Electric. High-pressure oil is used, controlled by the operator carried in the machine.

A severe problem is encountered with hydraulic drive mechanisms when operation is required in very low ambient temperatures, since the viscosity of the fluid is temperature-dependent and the fluid can even freeze at very low temperatures. Consequently, a special preheating procedure is necessary when starting-up at low ambient temperature. For example, with an ambient temperature as low as $-40°C$ a heating cycle, involving running the hydraulic power unit for 30 min before use, can be required.

Despite the problems, hydraulically powered artificial arms have been produced for prosthetic use[13]. The advantages claimed compared with an electromechanical system include fewer parts and less maintenance, improved operating characteristics in wet or dusty environments, adaptability to modular concepts, high power transmission efficiency and possibility of use of a single centralised power unit. A small electrically driven pump has been used, taking 0.42 A on no-load and a maximum of 1.2 A from a 12 V d.c. supply. The fluid used is octoil-S, a light diffusion pump oil with very little odour, supplied at a rate of 200 cm³/min at pressures between 800 kN/m² and 1600 kN/m² by a power unit weighing 0.3 kg and measuring $7 \times 7 \times 3$ cm, not including the supply batteries. The shoulder, the elbow, the wrist and the hand are all movable, the fastest motion being that of opening or closing the hand, which takes 1.0 s.

A practical portable hydraulic power supply

Typical of the work on portable power supplies using hydraulics for prosthetic use is that described by Davies[27]. A Lucas radial pintle form of pump was used, giving a peak efficiency of 78% at a pressure of 2800 kN/m², being driven at 2500 rev/min from an electric motor running at 7500 rev/min via reduction gearing. The gearing efficiency was 85%. A hydraulic accumulator was used in order to permit a high peak torque of 11 Nm at up to 3 rad/s rotation with a total maximum flexion of 150°, for use in an artificial elbow flexion actuator. A flexible bellows system made from PVC was used to prevent leakage of hydraulic fluid into the environment. The use of refrigerant gases such as Freon or possibly sulphur hexafluoride has been considered as a method of energy storage by making use partially of the liquid phase of the gas.

There are various problems to be considered in such an application of hydraulics, such as the possibility of cavitation; the fact that low pressure and high flow rate give a high pump efficiency but also a large size and weight of actuator, a larger volume of oil and a larger accumulator; and also the problems introduced by the entrapping of air in the hydraulic fluid.

The power pack proposed by Davies[35] can give 4.1 cm^3 per second of oil at a pressure of 2500–3000 kN/m^2 and a power of 10–12 W with an over-all efficiency of 45%. It can fill the accumulator in 12 s with enough fluid for two cycles of elbow movement. The weight is 0.56 kg, the 50 cm^3 accumulator weighing an additional 0.35 kg.

It has been shown by McLeish and Marsh[29] that while the weight of the fluid in a hydraulic system falls as the pressure used is increased, the weight of the accumulator increases. There is consequently an optimum value of pressure to be used for minimum weight; this optimum value lies at the impracticably high figure of some 10 000 kN/m^2. The minimum on the curve is very flat, but the curve rises steeply as the weight of fluid required increases below some 5000 kN/m^2.

McLeish and Marsh have considered methods of obtaining a low weight and a constant hydraulic pressure by the use of a storage cylinder formed from a rubber flexible tube inside a rigid constraining tube, though only low pressures have been obtained in this way. They have also suggested the possible use of the aerosol principle with the liquefiable gas Arcton 13 on one side of a flexible bag accumulator, though the control of temperature is important if this principle is to be used.

Batteries for electrical supply[4, 5, 7, 32, 42, 43]

In any particular application where it is required to use electrical batteries to supply mobile equipment it is first necessary to decide whether rechargeable cells or accumulators should be used or if instead dry cells can be used. If the primary requirement is a low running cost, then there is no doubt at all that rechargeable cells should be adopted. Although the initial cost of the rechargeable cell is probably some 4–20 times that of the equivalent dry cell, the total annual running cost using dry cells is likely to be between 4 and 30 times the annual cost of the rechargeable cell. Added to this there is the additional inconvenience of frequent purchase and replacement of dry cells, compared with the regular charging and

84 Power for Actuation

annual replacement of the rechargeable cell. In some applications, however, where the weight and the volume must be minimised, it is important that dry cells have only about one-half of the weight and volume of the equivalent rechargeable cells.

For power drive applications on robots the weight of dry cells would be excessive, and rechargeable cells are to be preferred. These should preferably not be of the lead–acid type, again because of the weight. Although the silver-zinc form of battery can show a weight advantage in many applications, the number of recharging cycles is restricted[12]. In addition, batteries of this type are not completely sealed and there is therefore some possibility of the escape of caustic alkali through the vent holes in unusual circumstances. This could happen if, for example, the robot was overturned.

Consequently, for many drive applications at the present time the superior safety, reliability and robustness of the nickel–cadmium cell make it preferable to the silver-zinc cell despite the doubling of weight which must be accepted.

It is not possible to quote accurate performance figures, since these depend very much on the particular application considered. However, a typical nickel–cadmium battery supplying a 12 V d.c. motor with a starting surrent of 5 A and a running current of 1.5 A and providing a total energy of 6 Wh/day will have a weight of about 0.7 kg and a volume of about 200 cm^3.

The capacity of such a battery would be about 14 Wh, so that in an emergency it could be used for 2 days without charging. Such a battery would cost about £12 per year, assuming that it was replaced annually, and the cost of charging would be negligible once the initial cost of the charging equipment had been met. In the writer's work it has been found advantageous to build simple charging devices into mobile robots, so that they can be charged directly from the a.c. mains without separate charging equipment.

In comparison, a 120 g cylinder of carbon dioxide gas has about the same weight, about twice the volume, about one-quarter of the energy capacity (and consequently only about one day's use per cylinder), and the running cost is about six times that of the annual running cost of the nickel cadmium battery system. However, for some applications the relative simplicity of pneumatic actuators makes the CO_2 system desirable.

Recent developments[12, 32, 36, 44–46]

More recently developed dry cells, of the mercuric oxide type, have a reduced weight and volume compared with earlier dry cells, so that the weight and volume are of the order of one-quarter of the equivalent nickel–cadmium battery. However, the running cost when such cells are used is even more excessive than with the earlier cells and the annual cost is some 50 times that of the equivalent nickel–cadmium battery. In addition, there is the usual problem of obtaining and storing replacement cells. Consequently, at the present time the nickel–cadmium cell is still to be preferred as a primary supply for robot energy.

It may be that in the future hydrogen–oxygen fuel cells[6, 24, 25] might prove to be an attractive source of power for robot applications. Such devices have been used in outer space applications, though perhaps here there is a weight advantage which is important. For domestic use the disadvantage of the fuel cell for the primary energy source is the need for compressed hydrogen. The fuel cell appears at present to have advantages for use in applications where power is required continuously for long periods, more than, say, 20 h continuous use. Provided that high pressures (4000 kN/m^2) and fairly high temperatures (200°C) are used, very high values of energy density, expressed in watt-hours per kilogram of weight, are obtainable with fuel cells.

One problem for certain applications of rechargeable cells is the electrolyte freezing point. This can be as high as $-27°C$ for an alkali cell.

In some prosthetic equipment it has been found advantageous to provide equipment to switch off the motors if the battery voltage falls to an excessively low value.

For some applications the so-called mechanically rechargeable battery can be used as a power source. In this form of battery the complete anode can be replaced whenever it has been used up. To do this it is only necessary to fit a new inner container in which there are a fresh anode and some fresh solute in solid form. The container is inserted into the water-filled cathode box. The time required for this replacement procedure can be much less than the recharging time of a storage battery.

The same cathode can be re-used for 50–100 replacements of the anode container, so that a long over-all effective life is obtained. The output from this form of battery can approach about 220 Wh of energy per kilogram of weight. The fact that the battery is usable over a very wide range of output currents makes it an attractive power source for robot use. In addition, this form of battery can operate under a wide range of environmental conditions and over a temperature range of $-40°C$ to $+54°C$. Despite the necessity for anode replacement, it has been claimed that the operating cost is competitive with that of other battery systems. Whether or not this form of supply becomes significant for robot use remains to be seen. It might be more convenient to replace the whole battery, and then possibly to reconstitute it later.

Another recent innovation has been the so-called maintenance-free battery, in which the reliability has been improved by the elimination of antimony and by the use of little free electrolyte. The life is about 3 years.

Some robots will be required to operate in a high ambient temperature, and in such cases the lithium–tellurium cell is attractive. This operates at a temperature of 500°C, and it will not supply power below 400°C. However, a cell weighing only 500 g is capable of giving a power output of 64 W, with a capacity of 100 W, while the volume is less than 500 cm^3.

The electrical capacity of a storage battery falls as the temperature falls, though under normal circumstances the specific gravity does not fall enough to allow freezing of the electrolyte. The fall of capacity can be as much as 15% for a 20°C decrease of temperature. Such problems cause difficulties when a battery-operated vehicle has to be used in a cold store. The difficulties are even more pronounced when a battery must be used in deep space conditions.

A guide to the over-all power requirements of the robot can be obtained by considering the equivalent power consumption of the human when carrying out various tasks. This varies from 100 W in light activity to over 1 kW in climbing stairs. The over-all efficiency of early robots can be expected to be lower than that of a human, and allowance must be made for this in deciding on the power requirements.

In some cases it is possible to derive the power for charging robot batteries from sunlight, by the use of solar cells. This method has proved particularly useful in space applications of robot devices. Radioactive batteries are a possibility for low powers.

The Japanese Matsushita Company has produced a new high-energy primary battery having an energy density some four to five times that of conventional calls. An organic electrolyte is used with a lithium negative electrode and a polycarbon positive electrode, the voltage being 2.6 V per cell. A sealed high-energy battery giving some five times the conventional power-to-weight ratios has been used to power a Chevrolet Vega car, which could run at 100 km/h and could cover 160 km before re-charging was required[33].

It can be seen that there are great possibilities of development in both primary and secondary forms of electrical power storage for robot use. For many purposes the secondary form is likely to be preferred, with the robot eventually taking care of its own charging arrangements from the mains.

Battery maintenance[7, 14, 17, 30, 34, 37, 38, 47]

Although the electrical storage battery appears to be a good choice for the main power source for a mobile robot, it must be remembered that there can be a number of problems which are by no means minimised when the battery has to be used in a domestic environment. For example, the electrolyte of a storage battery can be dangerous, though totally enclosed batteries such as the sealed nickel-cadmium cell are available.

The robot battery must be charged regularly. Charging can be thought of as the feeding of the robot, and as with the human or with any animal there is an optimum value of charging—not too little, not too much. Because of this, it is important to make this charging operation as automatic as possible and to arrange for the robot to supervise and arrange its own charging as far as possible.

The initial charging of the battery is most important and should preferably be carried out at the factory. The rate of charging at the start of a routine charge can be arranged to be quite high, a typical figure having been quoted as 7% of the ampere-hour capacity of the battery (e.g. 7 A for a 100 Ah battery) However, if there is more time available for charging, the rate should be lower, and it should in any case be reduced as the charge progresses.

As the charge continues, the voltage of the battery will gradually rise, and it is useful to control the charge by reducing the charging current as soon as the voltage reaches certain limits. There are various methods of charging control, each of which has its supporters. It is probably best to consult the battery manufacturers for their recommendations, since they have most to lose in the long run if an unsuitable method is adopted. One recent possibility is the use of gas control, in which the heat produced by catalytic recombination of the oxygen and hydrogen given off during gassing is used to control the rate of charging. The fact that such gases are produced can itself constitute a danger to the domestic environment because of the danger of explosion.

The general requirements of a battery for robot use are very similar to those for any other mobile use, namely long life, little deterioration when the battery stands idle, good mechanical strength and resistance to the effects of vibration and shock, and good resistance to the effects of overcharging. For many applications a wide range of permissible temperatures is also a useful feature. In general, a minimum weight, size and, of course, cost are also required. Obviously, the ideal battery has not yet been produced but progress towards it is most important to the whole future of robotics.

Sometimes, in order to make the optimum use of capital

equipment, the robot will be require to operate 24 hours each day except for short routine inspection and maintenance. In such cases the supply batteries must be interchangeable and easily replaceable.

Rapid charging of storage cells[8, 9, 10]

In some applications of robots it will be important to be able to recharge the electrical storage cells quickly. This will be particularly so in the early days of robot production, while the robot is still relatively expensive and therefore in short supply. It is, of course, possible to change the cells of a robot so that the robot can continue to work even while the first set of batteries are being charged. However, some of the recent work on rapid charging of storage cells makes it appear that it might be quite unnecessary to change the batteries provided that a short rest period can be allowed for 'feeding'.

When a battery is fully charged, gassing starts freely. It is possible to detect the resulting increase of gas pressure and to use this to terminate or to slow down the rate of charging. The increase of cell temperature can be used in a similar way. Other methods of control which could possibly be used are based on the measurement of cell voltage or on the use of the rather expensive coulometer.

One method which has been used with success is based on the current-voltage characteristic of the charging device. This is arranged to charge at a rate of $35 \times I_s$ (where I_s = nominal capacity/10 h) until the voltage reaches a certain maximum value. The charge is then continued at trickle charge rate of $I_s/5$ until the voltage falls to a certain minimum value, when charging is continued at the higher rate. By this means it is possible to charge in only 20 min instead of the usual 14 h.

REFERENCES

1. Fyson, J., *et al.*, 'Design Considerations in a Myoelectric Hand Prosthesis', *Proc. IEE*, **116**, February, 281 (1969).
2. Anon., 'Machine Clomps across Country', *Electro-Technology, N.Y.*, **8**, May, 29 (1969).
3. Blake, L. R., 'Conduction and Induction Pumps for Liquid Metals', *Proc. IEE*, **104A**, July, 49 (1956).
4. Barak, M., 'Batteries and Fuel Cells', *Proc. IEE*, **117**, 1561 (1970).
5. Whiteway, F. E., 'Power Sources in Bioelectrical Engineering', *Electron. & Power*, **16**, 309 (1970).
6. Lindstrom, O., 'The Fuel Cell now makes its Entry on the Market', *A.S.E.A. Jl*, **40**, 91 (1967).

7. Smith, G., *Storage Batteries*, Pitman (1964).
8. Ball, J. V., 'There Is No Overcharge for Fast-Charged Batteries', *Electronics*, **41**, January 22, 97 (1968).
9. Zinder, D. A., 'Fast Charging Systems for NiCd Batteries', *Trans. IEEE*, **IGA4**, September/October, 555 (1968).
10. Lin, W. C., et al., 'Fast Charging Circuit for NiCd Battery Used in Implant Electronic Systems', *Trans. IEEE*, **BME17**, October, 331 (1970).
11. Pech, J. F., 'Electric Clutches', *Mach. Des.*, **41**, December 18, 47 (1969).
12. Klein, P. M. V. and Montgomery, S. R., 'Portable Power Supplies for Prostheses', *Proc. Instn Mech. Engrs*, **183**, pt 3J, November, 26 (1968).
13. Stevenson, D. A. and Lippay, A. L., 'Hydraulic Powered Arm System', *Proc. Instn Mech. Engrs*, **183**, pt 3J, 37 (1968).
14. Fryer, T. B. and Sandler, H., 'A Rechargable Battery System', *Proc. 23rd Conf. Eng. Med. Biol.*, Vol. 12 (1970).
15. Davies, R. M. and Lambert, T. H., 'The Stability of Pneumatically Powered Prostheses', *Proc. 23rd Conf. Eng. Med. Biol.*, Vol. 12, 244 (1970).
16. West, J. C., 'Clutches for Control Systems', *Control*, **12**, January, 48 (1968).
17. Sauer, H., 'A Semiconductor Charging Device for Auxiliary Batteries on Short Distance Transport Vehicles', *Siemens Rev.*, **35**, 313 (1968).
18. Seabors, G. T., 'The Eligible Kinsfolk of Uranium', *New Scient.*, **38**, May 23, 410 (1968).
19. Wilson, A. B. K., 'Recent Advances in the Control of Externally Powered Artificial Limbs', *Proc. Instn Mech. Engrs*, **179**, pt 3H (1964).
20. Montgomery, S. R., 'Design of an Experimental Arm Prosthesis: Engineering Aspects', *Proc. Instn Mech. Engrs*, **183**, pt 3J, November, 68 (1968).
21. Metzger, J., 'Hydraulic or Pneumatic?', *Mach. Des.*, **41**, June 26, 126 (1969).
22. Boulden, L. L., 'Hydraulics, Pneumatics, and Dollars', *Mach. Des.*, **43**, February 18, 104 (1971).
23. Brann, R. P., 'Linear Analysis of a Pneumatic Differential Actuator with Positive Feedback for Prosthetic Control', *Control*, **10**, August, 428 (1966).
24. Young, G. J., *Fuel Cells*, Reinhold (1963).
25. Liebhafsky, H. A. and Cairns, E. J., *Fuel Cells and Fuel Batteries*, Wiley (1968).
26. Hall M. J. and Lambert, T. H., 'Artificial Limbs—An Engineering Appraisal', *Hosp. Engr*, **19**, February (1965).
27. Davies, B. L., 'A Prototype Portable Hydraulic Power Supply for Prosthetic Applications', *Proc. Conf. Hum. Locomotor Eng., Sussex, September, 1971* (Instn Mech. Engrs), 285.
28. Cool, J. C. and Pistecky, P. V., 'Miniature Electrically Operated Proportional Valve', *Proc. Conf. Hum. Locomotor Eng., Sussex, September, 1971* (Instn Mech. Engrs), 271.
29. McLeish, R. D. and Marsh, J. F. D., 'Hydraulic Power from the Heel', *Proc. Conf. Hum. Locomotor Eng., Sussex, September, 1971* (Instn Mech. Engrs), 211.
30. Robertson, J., 'What is there to Choose between Traction Battery Chargers?' *Mech. Handl.*, **58**, October, 95 (1971).
31. Lord, M. and Chitty, A., 'Stabilisation of Pneumatic Prosthetic Systems', *Proc. Conf. Hum. Locomotor Eng., Sussex, September, 1971* (Instn Mech. Engrs), 307.
32. Barak, M., 'Developments in Electrochemical Energy-Conversion Devices. Batteries and Fuel Cells', *Proc. IEE*, **112**, July, 1439 (1965).
33. Anon., 'High-Energy Battery Promising for Electric Cars', *Electl Rev.*, **190**, May 26, 720 (1972).

34. Soderholm, L. G., 'Voltage Sensing Circuit Matches Charge to Battery Condition', *Des. News*, **27**, March 20, 46 (1972).
35. Davies, B. L., 'A Portable Hydraulic Power Unit for Prostheses', *Eng. Med.*, **1**, January, 41 (1972).
36. Salihi, J. T., 'Two for the Road', *Spectrum*, **9**, July, 43 (1972).
37. Moreton, P. L., 'Fuel Gauge for the Electric Car', *Proc. IEE*, **119**, June, 649 (1972).
38. Brugger, C., 'Industrial Battery Charger for Mobile Equipment', *Control & Instrum.*, **4**, October, 45 (1972).
39. Donald, A., 'Development of a Control System for a Single Digit Artificial Limb', University of Aston (1970).
40. Younkin, G., 'Which Drive, Electrical or Hydraulic?', *Control Eng.*, **19**, November, 50 (1972).
41. Anon., 'Don't Let Your Fluids Be a Drain on Profits', *Metalworking Prodn*, **117**, March, 60 (1973).
42. Collins, D. H., *Power Sources*, Pergamon, (1970).
43. Feder, D. O. and Biagetti, R. V., 'The Lead-Acid Battery, a New Shape for an Old Workhorse', *Bell Labs Rec.*, **50**, August, 207 (1972).
44. Heise, G. W. and Carey, C. N., *The Primary Battery*, Wiley (1971).
45. Barak, M., Recent Developments in Batteries and Voltaic Cells', *Electron. Power*, **18**, August, 290 (1972).
46. Anon., 'A New Electric Vehicle Project', *Original Equipt Mfg Des.*, **2**, March, 72 (1973).
47. Wilke, W., Voltage Monitor Protects NiCd Batteries, *Electronics*, **46**, March 1, 85 (1973).

6

Robot stability

Introduction

The subject of the possible methods of stabilisation of regulator and servo systems is, of course, a vast field of study. Nevertheless, a great deal of the published information in this field deals with purely theoretical methods which have been little used in the stabilisation of practical devices.

In these circumstances, simplified design techniques, using assumptions of some linearity, and giving a starting point in the design process, are found to be most useful[24]. In this chapter miscellaneous investigations which have been carried out, and methods which have been used by, the writer in industry, in consultancy and in university laboratory will be described briefly. The treatment cannot be exhaustive but the approaches described will be found to be useful in practice.

Simplified feedback stabilisation

In order to meet a complex specification for a closed-loop control system, advanced and lengthy design techniques are often required. However, in some cases performance specifications are less severe, and it can be assumed as a first approximation that the system is linear. It can then be possible to adopt simplified methods of design.

For example, in a wide range of problems the performance is not critical and the main object of design is to ensure that a specified accuracy can be obtained without instability. Occasionally, very tight specifications are laid down which would be very difficult and consequently very expensive to meet. In the writer's experience such

Robot Stability

tight specifications often prove on enquiry to be completely unnecessary in the intended application.

Usually in cases where the performance requirements are not specific and critical the major components of the control loop are specified in advance and the necessary over-all gains follow directly from the minimum acceptable performance. Provided that the required design information on the major components is available (and unfortunately this is very often not the case), the main problem remaining is then the achievement of stability.

It is desirable to have some method of rapid, if rough, design of a stable control loop, even if only to be able to estimate a price for a proposed control scheme without undue delay. The approach to be described here gives a simplified way of thinking about feedback stabilisation.

Stabilisation of regulator systems[1]

In order to introduce the method, it will help to consider first one approach to the reshaping of transfer functions by the cascade addition of an equalisation network. If we have an open-loop system giving a transfer function

$$\frac{V_x}{V_1} = \frac{A_1}{(1+sT_1)(1+sT_2)}$$

and this is cascaded with a phase advance network having a transfer function

$$\frac{V_0}{V_x} = \frac{A_x(1+sT_3)}{(1+sT_4)}$$

then the over-all transfer function becomes

$$\frac{V_0}{V_1} = \frac{A_1 A_x(1+sT_3)}{(1+sT_1)(1+sT_2)(1+sT_4)}$$

Now if $T_3 = T_2$, and this can easily be achieved by the correct design of the phase advance circuit, then the terms involving these time constants are cancelled out and the over-all transfer function becomes

$$\frac{V_0}{V_t} = \frac{A_1 A_x}{(1+sT_1)(1+sT_4)}$$

Robot Stability

It can be seen here that the phase advance network has been effectively used to replace T_2 in the original transfer function by T_4 in the new transfer function, so reshaping the response. The process is well-known, and it is widely used to obtain an improvement in the transfer function. The simplicity of the process leads one to wonder if it is possible to find similar techniques that will be of use where feedback stabilisation is required to be used—for example, in order to provide some allowance for unavoidable characteristic drifts.

Feedback stabilisation

If a system is to be designed with feedback stabilisation as shown in the block diagram of *Figure 6.1*, it is quite usual to consider the stability of the system by calculation of the open-loop response obtained by imagining the loop to be opened at point X. Then the open-loop response is

$$\frac{V_f}{e_x} = \frac{A_1 A_2 A_3 A_f(1+sT_4)}{(1+sT_2)(1+sT_3)(1+sT_4) + A_2 A_3 A_4 sT_4} \times \frac{1}{(1+sT_1)}$$

It can be seen here that if the design ensures that $T_4 = T_1$, then one term is cancelled out on top and bottom.

However, unless the denominator can easily be factorised, the problem remains of putting this open-loop response into some form in which it is possible to see the effect on the over-all closed-loop stability of the system of any variations of A_4 and possibly also of T_4. There are various possible approaches, and these will not be elaborated here since they have been well covered in the literature on servo systems.

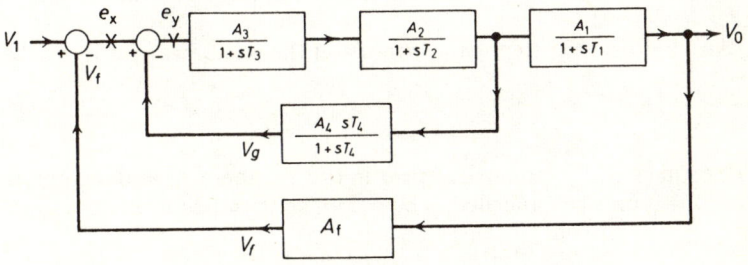

Figure 6.1 Feedback stabilisation block diagram

The primary interest here is in the closed-loop stability of the whole system. In order to study this it is not at all essential to break

94 Robot Stability

the feedback loop, or to imagine it to be broken, at the point X in *Figure 6.1*. The loop can instead be imagined to be broken at point Y. It is then instructive to redraw *Figure 6.1* as shown in *Figure 6.2*.

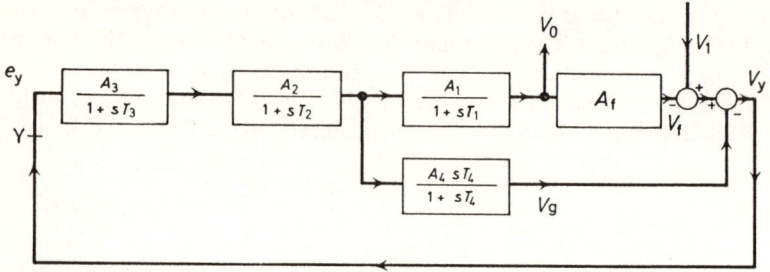

Figure 6.2 Development of the circuit in Figure 6.1

The two diagrams are identical in function, the only difference being in the way that they are drawn. The feedback which was around T_2 and T_3 is now shown instead as a feed-forward around T_1, the main feedback being via a unity-gain path. Now if the loop is broken at point Y, the open-loop response becomes

$$\frac{V_y}{e_y} = \frac{A_2 A_3}{(1+sT_2)(1+sT_3)} \times \left(\frac{A_1 Ai}{1+sT_1} + \frac{A_4 sT_4}{1+sT_4} \right)$$

This response can now be shaped correctly to avoid instability when the loop is closed.

The first simplification can be achieved by making $T_4 = T_1$ as mentioned above. The equation then becomes

$$\frac{V_y}{e_y} = \frac{A_1 A_2 A_3 A_f(1+s(T_1 A_4/A_1 A_f))}{(1+sT_1)(1+sT_2)(1+sT_3)}$$

This shows that the control loop can be simplified to that of *Figure 6.3*.

An obvious next step is to proportion the feedback gain A_4 so that

$$\frac{T_1 A_4}{A_1 A_f} = T_2$$

The term $(1+sT_2)$ then will appear in both numerator and denominator, and it can be cancelled. The response then becomes

$$\frac{V_y}{e_y} = \frac{A_1 A_2 A_3 A_f}{(1+sT_1)(1+sT_3)}$$

Now here there are only two time constants involved and, consequently, instability is inherently impossible.

Robot Stability 95

In effect, T_2 has been eliminated from the open-loop equations by following the stabilisation procedure here. Consequently, the open-loop time constants have been reduced in number from three, which can easily cause instability, to two, which can never be unstable.

Fairly lengthy derivation of the closed-loop response of the system gives

$$\frac{V_0}{V_1} = \frac{A_1 A_2 A_3}{(1+sT_1)(1+sT_3) + A_1 A_2 A_3 A_f} \times \frac{1}{1+sT_2}$$

Now the first part of the expression on the right-hand side here is the closed-loop response of

$$\frac{V_y}{e_y} = \frac{A_1 A_2 A_3 A_f}{(1+sT_1)(1+sT_3)}$$

where the A_f part is regarded as being in the feedback section. The second part of the expression for the closed-loop response is simply the response of the time constant T_2. Thus following the simplified

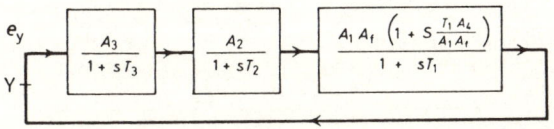

Figure 6.3 Simplification of the control loop

stabilisation procedure given here can be regarded as being equivalent simply to casting T_2 outside the control loop and so avoiding the possibility of instability.

In some practical cases it will not be permissible to have time constant T_2 effectively outside the control loop, since the resulting over-all closed-loop performance will not be acceptable. However, even in such cases the procedure is such a simple and rapid one that it has been found to be a good starting point. In many applications time constant T_2 is short and its appearance outside the loop is not objectionable. The procedure can easily be developed for use in those cases where there are more than three time constants in the loop. The procedure can be repeated if required until all, or until all but one, of the time constants have been effectively cast out of the loop. However, in practice it is not usually found to be necessary to reduce the number of time constants still effectively in the loop below two.

Robot Stability

Extension of simplified stabilisation

It is possible to extend the simplified method of stabilisation design to cases where there is an integrator included in the control loop. As an example, in the arrangement of *Figure 6.4*, time constant T_2 can be effectively cast out of the loop if $T_4 = T_1$, where feedback equals $A_4/(1+sT_4)$ and $A_4 = A_1 A_f T_2$. However, this procedure does not make the system inherently stable.

If instead

$$\text{feedback} = \frac{A_4(1+sT_4)}{(1+sT_1)}$$

and

$$A_4 = A_1 A_f (T_2 + T_3) \quad \text{and} \quad T_4 = \frac{T_2 T_3}{T_2 + T_3}$$

then the open-loop response becomes simply

$$\frac{A_1 A_2 A_3 A_f}{s(1+sT_1)}$$

and instability is impossible.

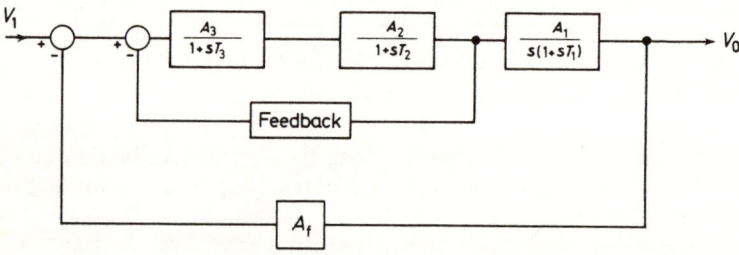

Figure 6.4   Control loop with integrator

If now the suggested feedback parameters are inserted and the resulting closed-loop response is laboriously computed, it is found to be

$$\frac{V_0}{V_1} = \frac{1}{(1+sT_2)(1+sT_3)} \times \frac{A_1 A_2 A_3}{s(1+sT_1)} \bigg/ 1 + \frac{A_1 A_2 A_3 A_f}{s(1+sT_1)}$$

This closed-loop response corresponds to that which would be obtained with a system of the form of *Figure 6.5*.

Thus both of the minor time constants have been in effect cast out of the loop and instability is rendered impossible.

The transient response obtained by following this very simple design procedure may not be directly acceptable for some applications, but the procedure is so quick and simple that it is always a useful starting point.

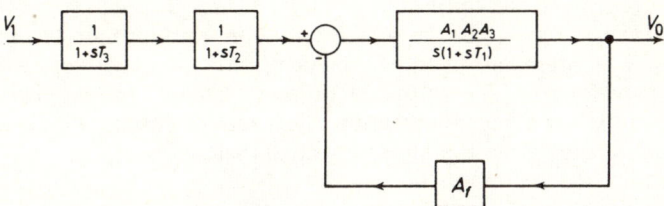

Figure 6.5   Closed-loop response with feedback parameters

It is possible to extend the method for use with nonlinear systems. In either linear or nonlinear systems it is vitally necessary to investigate the effects of variation of parameters if a reliable system is to be obtained. In such a study it is very useful to be able to compare the root values of quadratic equations appearing in the denominator with those of quadratic equations appearing in the numerator.

The writer has shown[2] how the roots of the quadratic equation $x^2 + ax + b^2 = 0$ can be expressed as $-b \exp(\pm jL)$, where $\cos L = a/2b$ ($a/2$ less than b), or $-b \exp(\pm L)$, where $\cosh L = a/2b$ ($a/2$ greater than b), depending on the value of $a/2$ compared with b. This result is very easy to visualise geometrically, the path of the roots as the value of $a/2$ is varied being along a hyperbola, leading into a circle. If the roots are real, their ratio is simply $\exp 2L$. These results have been found most useful in the practical design of servo control systems. The writer has used the approaches described here in the design of some successful systems[3] which were so nonlinear as to be virtually undesignable by conventional techniques, particularly where the effective time constants are known to vary over a very wide range.

There are at the present time many possible methods of design of control systems under investigation all over the world[22]. No single approach yet appears to have met with universal application and approval, and few of these methods have received practical verification. Nightingale and Todd[25] have considered the preliminary design of prosthetic devices having adaptive control systems, the principle adopted being that human unconscious actions should be carried out automatically by the machine, while those actions which are normally carried out consciously by the human should retain conscious human control in the prosthetic device[26].

In all cases it is necessary for the designer to set a target for the control system to achieve. Once this is done, it is possible to construct a control system which can adapt its own parameters so as to achieve the target control function as closely as possible, and if conditions change it will then readjust its control system in order again to achieve the optimum.

However, it must be recognised that in the animal system the optimum is often only vaguely defined, and it is often necessary to anticipate far into the future in order to achieve the correct conditions now for a future optimum. This sort of anticipation has not yet been achieved with adaptive control systems.

Electronic control of series motors

Until recently it was usual for the series motors used for traction applications to have a stepped form of speed control. With the development of the thyristor, however, it has become possible to fit such motors, operating from a battery supply carried with the vehicle, with a continuously variable speed control[4].

Typically, one thyristor is used to cause current to flow from the battery through the motor. While the current is flowing, a capacitor is charged up. After a short time, a second thyristor is fired and this is arranged to discharge the capacitor through the first thyristor in such a direction that its conduction is terminated. A 'flywheel' form of rectifier is connected across the motor armature so that the motor current can continue to flow even after the thyristor has disconnected it from the supply. The speed can be controlled by variation of the frequency of switching of the thyristors.

There has recently been extensive work on the use of such arrangements for the control of traction devices such as fork-lift trucks, and this work is likely to be of direct application to the control arrangements of the mobile robot.

Field control of motors

It is usual to think of the control of electric motors for traction purposes as necessarily being of the series motor control type. Methods of control of series motors are quite well known from their use in electric vehicles and also in cranes and hoists. However, there is another method, more recently developed, which can give a constant torque form of characteristic and which yet requires only control of the relatively low-powered field of the electric motor.

The method is based on the well-known arrangement in which a low-voltage motor has its armature supplied via a high value of resistance from a high-voltage supply[5-11]. However, instead of using a resistor, which introduces a quite high power loss and therefore limits the permissible motor power level, the new method uses a relatively low loss arrangement.

The Boucherot circuit[12, 13] comprises a series tuned circuit supplied at its resonant frequency from an alternating supply. Across one of the components, either the inductor or the capacitor, is connected the load. In the present case[14] the load consists of a rectifier which supplies the armature of the motor (*Figure 6.6*).

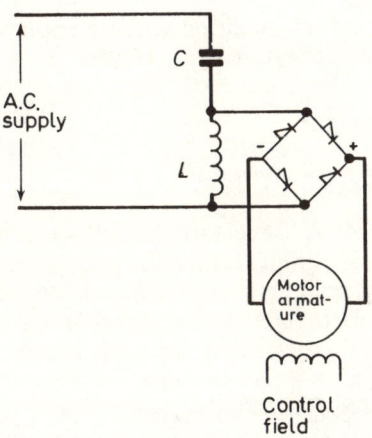

Figure 6.6   Boucherot circuit in constant torque drive

It is a property of the Boucherot circuit that it produces a constant current in the load, regardless of its resistance or of its back-e.m.f. Because of this constant current form of supply to the motor armature, it is quite possible to remove the field excitation of the motor completely without taking an excessive armature current from the supply. If, however, a field current is supplied, then the armature produces a constant torque, the value and direction of the torque depending directly on the value and direction of the field current of the motor. The torque is almost independent of the speed of rotation of the motor. Such a characteristic is ideal either for robot traction purposes or for robot lifting mechanisms.

This form of supply has the advantage that since there need be little voltage lost in the constant current circuit, the motor voltage rating can be approximately the same as the supply voltage value, and it is not necessary to use a low-voltage motor on a high-voltage supply.

The disadvantage of the system is the cost and the weight of the resonant capacitor and inductor required in the Boucherot circuit. The fact that the supply to the Boucherot circuit must be alternating is no great disadvantage these days, since inversion techniques are well known. However, once again engineering economics must be considered.

When compared with the well-known Ward Leonard system of control, it is found that the over-all cost is not much less. However, the real advantage for robot applications is the possibility of reduced size and weight which also makes the system attractive for normal hoist purposes.

The system was investigated for possible application to cybernetic and robot controls at Aston by J. Clarke, R. C. G. Massey and L. K. Banks[14].

Control of double-stator squirrel cage motor[23]

The engineer P. M. J. Boucherot is well known for his form of double-cage electric motor, which is widely used. His important work on the Boucherot effect, and on the earlier forms of his motors, is less well known. In fact, Boucherot himself wrote little about these motors. Nevertheless, they have the possibility of application to robotics as variable-speed machines, and they have been investigated at Aston by B. Earle and A. Robertson[23].

The double-stator machine has a single squirrel cage rotor, with two separate stators side by side axially with the shaft of the rotor. It is found that a considerable variation of the torque–speed characteristic of such a machine can be obtained in various ways.

If both of the stator windings are energised from the same polyphase supply, then physical rotation of one of the stators about the common axis and relative to the other stator has the effect of varying the relative e.m.f.s induced in the two ends of the cage rotor and so of varying the circulating current in the rotor and, hence, the torque produced. An alternative method of obtaining variation is by varying the actual value of the voltage applied to the second stator, while keeping the relative positions of the two stators fixed and the voltage amplitude applied to one of the stator windings fixed. Alternatively, the phase of the voltage applied to the second stator can be varied with respect to the voltage applied to the first stator. The amplitude or the phase can be varied electronically, if required. Although the technique is promising, the special motors are unobtainable.

One problem encountered in the experimental work on this technique was ferroresonance[15-17], caused when capacitors were added to the circuit to tune out some of the effective inductive impedance of the windings.

Simple mobile machines[18-22]

In order to be able to investigate problems or reliability in mobile robots at Aston, the performance of small 'tortoise' machines is being investigated. Early models of such machines had a poor reliability, and it has been possible to improve this by a study of the reasons for each failure.

The main points of interest in such a study are:

1. The operation of electronic control and logic circuits on a battery supply.
2. The necessary size of batteries and the required frequency of recharging and rate of recharging.
3. The interaction between different parts of the control circuitry when operated on common supplies.
4. The possibility of arranging the robot to plug itself into a charging supply whenever necessary. (This can fancifully be thought of as 'feeding'.)
5. The effect of a mobile environment on the control equipment and on the moving parts of the device.
6. Problems of cleanliness in a mobile environment.
7. The required speed of movement.
8. Methods of conveying to the mobile robot information about its environment.
9. Efficient methods of control of the motive power.
10. Safety factors such as obstacle-detecting switches, emergency stops and their locations on the robot.
11. Rapid and efficient methods of dismantling for maintenance which at the same time are 'child-proof'.
12. Possibilities of redundancy of parts and the acceptance of reduced performance in emergencies. Is it preferable to stop in the event of any failure or to continue operation despite a reduced performance?
13. Possibilities of self-repair.
14. Noise produced by the robot.

It can be seen from this list, which is not at all a complete one, that study of small mobile robots can help to clear up some of the problems even before they are encountered on full-size, mobile, working robots. Some of the problems can in fact be studied on non-mobile equipment, but a final study on mobile equipment is desirable.

The simple mobile machine Astor

Several experimenters have used the form of the beetle or of the tortoise for simple mobile robot machines. Perhaps one of the best publicised has been that due to Walter[18]. This form was later elaborated by Angyan[19].

In the Aston Cybernetics Laboratory experiments have been carried out with the small machine Astor, which also has the general shape of a tortoise[22]. The mechanical arrangement was first produced by P. A. Kidd and C. J. Lloyd, who adopted the use of transistors in the control system in an attempt to improve the reliability of the earlier forms.

The early model has three wheels, two at the rear and one centrally situated at the front. The rear wheels are free-running. The front wheel has two operating motors. One of these, the drive motor, causes the wheel to drive the model along. The second, the scan motor, causes the wheel to rotate about its vertical axis, so giving a steering action. All of the possible actions of the device are caused by the action of these two motors.

The movement of the device is influenced by inputs from two photocell 'eyes' at the front, by a microphone 'ear', and by a movable shell which can detect the presence of obstacles by touch.

The model is designated to operate as follows:

1. If there is no external stimulus, the steering operates in such a way that the device searches for a light source.
2. If any obstacle is encountered, the model backs away before resuming its search for a light source.
3. If a weak light is detected by one of the photocells at the front of the device, then the motors operate to drive in the direction of the light.
4. If a strong light is detected by the other photocell at the front of the device, then the motors operate to drive away from the light.
5. If the sound of a whistle is detected by the microphone, then the model stops completely, or 'freezes' for a time.
6. If a certain number of coincidences occur of weak light de-

tection and the detection of a whistle sound, then conditioning occurs and for a time the sounding of a whistle has exactly the same effect as the detection of a weak light source.

In order to achieve these modes of action, the Astor device is connected as follows:

1. To simulate a searching mode, both motors are energised simultaneously. The motor speeds are regulated to ensure that, although the model moves about in an apparently random fashion, a complete 360° arc of search is covered in time.
2. To avoid an obstacle, the drive motor is switched off for a short period whenever an obstacle is encountered. This gives the scan motor time to turn the wheel away from the direction of the obstacle before drive recommences.
3. Whenever the photocell at the front of the device detects a weak source of light, the drive motor is immediately switched off but the scan motor continues to operate until the drive wheel is aligned in the direction of the light. The scan is then switched off and the drive motor is reconnected so that the device moves in the direction of the light source. There is thus a waiting period while the steering is aligned.
4. If the light detected by the photocells becomes too strong, the scan motor is reconnected and the search for a moderate source of light restarts.
5. Both motors are switched off if a whistle is sounded, so causing the model to 'freeze'.
6. In order to simulate a conditioned reflex action, the weak light input and the sound input are taken to a coincidence detector, the output of which goes to a counter circuit. The counter gives an output only after the weak light and the whistling sound have coincided a number of times. The output then gradually fades, if it is not reinforced, to simulate the forgetting feature. While the coincidence memory output is present, any further soundings of the whistle are temporarily arranged to act as though an input from a weak light source had been received.

The over-all block diagram of this control system is shown in *Figure 6.7*. This is composed mainly of integrated circuits and was designed and constructed by B. J. Alston and I. Foxall of the Cybernetics Laboratory at Aston, to replace an earlier scheme using discrete components[22].

The work on these schemes has revealed the sort of problems which are likely to be found with mobile robot devices. Later work has been based on this early work, and is aimed at the mounting of

104 Robot Stability

the Astra mark 3 machine on a mobile trolley, in order to investigate further the problems involved with the mobile and independent robot.

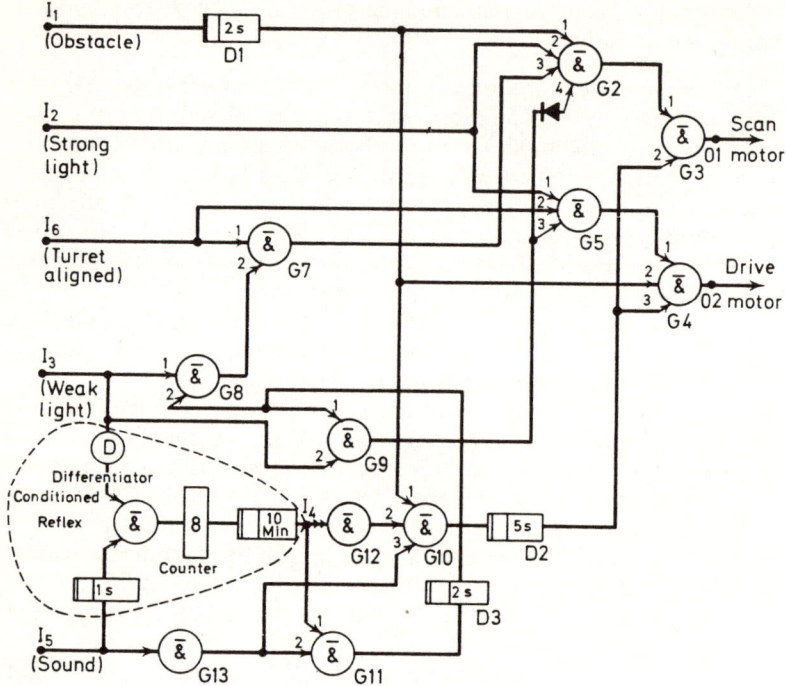

Figure 6.7 Logic diagram of Astor model

REFERENCES

1. Young, John F., 'Simplified Feedback Stabilisation', *Process Control Automn*, **10**, June, 233 (1963).
2. Young, John F., 'The Roots of Quadratic Equations', *Control*, **7**, November, 237 (1963).
3. Young, John F., 'Electronic and Magnetic Amplifier Voltage and Frequency Regulators', *Electricity in Industry*, No. 9 (1956).
4. Mazda, F. F., 'An Electric Vehicle Controller', *Electron. Components*, **12**, March 19, 235 (1971).
5. Andriesse, R. D., 'Split Field Servo Motors', *Control*, **9**, August, 425 (1965).
6. Anon., 'Characteristics of British Servo Motors', *Automn Prog.*, June, 198 (1960).
7. Williams, F. C., 'The Velodyne', in: *Servomechanisms*, H.M.S.O., 134 (1951).
8. Williams, F. C. and Uttley, Albert M., 'The Velodyne', *JIEE*, **93**, pt 3A, 1256 (1946).
9. Lampert, W. E. C., 'Naval Applications of Electrical Remote Position Controllers', *JIEE*, **94**, pt 2A, 236 (1947).

10. Taylor, P. L., *Servomechanisms*, Longman, 193 (1960).
11. West, J. C., *Servomechanisms*, English Universities Press, 182 (1955).
12. Young, John F., 'The Boucherot Effect', *Wireless Wld*, **68**, August, 391 (1962).
13. Young, John F., 'Voltage Regulator Comparison Circuits', *Control*, **6**, June, 90 (1963).
14. Young, John F., 'Field Control of Motors with Constant Current Armature Supply', *Control*, **12**, January, 35 (1968).
15. Young, John F., 'Ferroresonance; Problems and Aplications', *Electl Rev.*, **176**, May 21, 782 (1965).
16. Young, John F., Bibliography on Ferroresonance, deposited with Library at University of Aston.
17. Young, John F., 'Using Capacitors to Improve Low-Cost Magnetic Amplifiers', *Engineer, Lond.*, **220**, July 30, 176 (1965).
18. Walter, W. G., *The Living Brain*, Duckworth (1953).
19. Angyan, A. J., 'Machina Reproducatrix', in: *Mechanism of Thought Processes*, Vol. 2, 933, H.M.S.O. (1959).
20. Zemanek, H., et al., 'A Model for Neurophysiological Functions', in: Cherry, C. (ed.), *Information Theory, 1960*, 270, Butterworths (1961).
21. Kukhtenko, A. I., The Dynamics of Devices Which Imitate Living Organisms', in: Coales, J. F. (ed.), *Automatic and Remote Control*, Vol. 2, 658, Butterworths (1961).
22. Young, John F., *Cybernetics*, Iliffe (1969).
23. Young, John F., 'Control Possibilities of Double-Stator Squirrel-Cage Motors, *Control*, **12**, May, 416 (1968).
24. Young, John F., 'Phase Measurements in Feedback Amplifiers', *Electron. Eng.*, **27**, July, 311 (1955).
25. Nightingale, J. M. and Todd, R. W., 'Adaptive Control of a Multi-Degree of Freedom Hand Prosthesis', *Proc. Conf. Hum. Locomotor Eng., Sussex, September, 1971* (Instn Mech. Engrs), 249.
26. Nightingale, J. M., 'Intelligent Prostheses', *Proc. I.E.E. Meeting on Uses of Robots, London, April, 1973*, 3

7

Robot mobility

Introduction

The traditional robot of fiction is very humanoid and is consequently freely mobile. However, most present-day robots are completely fixed and can only be moved around with the intervention of a human. The exceptional robot devices which are mobile are generally restricted to fixed paths, either on railway lines or guided by buried conductors[1, 21]. Those fully mobile robots which have been produced have, in general, been under direct continuous human supervision. Examples are the robot devices which have been constructed for the handling of dangerous explosive or radioactive materials.

A serious limitation of some robot devices introduced for industrial purposes has been their lack of mobility. For example, it is too much to expect that an industrial machine loading robot device can maintain its action with complete absence of human supervision, unless it is simply built as a integral special-purpose part of a special-purpose machine. Even a slight movement of the nominally fixed position of the loader with reference to the machine being loaded will cause the parts to be incorrectly positioned, assuming that a general-purpose loading robot detached from the machine being loaded is in use.

The addition of some form of feedback control to the fixed-position loader can help to ease the situation, though this is rarely if ever done. One reason for this is that if the loader is to be fixed in position, then very long, very flexible and very controllable loading arms would be required. Also, some form of position sensor, preferably visual, would be required. It will be seen that, to cater for any unpredicted movements, a great deal of complication would have to be added to the simple machine loading robot.

One reason for this is the danger of damage in an industrial environment which might be caused by the use of mobile robot equipment in close proximity to the dangerous moving parts of high-production industrial machines. The possible damage to the machine tool is probably even more important than the possible damage to the mobile robot, both because of the great initial cost of modern machine tools and because of the very high cost of any down-time caused by the breakdown. A loading robot could be replaced by a human loader in an emergency. An expensive machine tool could not.

For such reasons it is likely that for some time mobile robots will be of several restricted types. Firstly, there is the guided robot for use in factories and warehouses as a means of transport and the rail restricted robot train. Secondly, the mobile robot is likely to be introduced into the domestic environment as a logical extension to the present non-mobile but fully automatic domestic machines. There is a third category, the supervised robot such as the autopilot for use on air or sea. Robots of this type are fully mobile, but they are not expected to work completely unsupervised by a human. A fourth type is semi-mobile in that it is coupled to a controlling computer by a trailing electrical lead. Another is the mobile farm machine[17, 18].

Perhaps the most spectacular example of the partly supervised robot has been the Russian robot device Luna 16 which visited the moon, collected samples of lunar dust and then returned to earth safely. This feat should not be underestimated, because even if the robot were actuated by a human, the inevitable delay between human command, reception by the robot, performance of the act and feedback to earth of the result of the performance makes this a very difficult task.

Advances are being made continuously in battery electric vehicles for passenger and for goods transport[29-31], and some of these advances will be applicable to the mobile robot. For example, in some cases electric motors have been built directly into the wheels of fork-lift trucks. Where the driven wheels are also used for steering, the manoeuvrability can be improved.

For such tasks as planetary exploration the independent mobile robot which can take orders from humans and which can report back its findings to humans, but is otherwise free, is a very attractive idea. The Russian Lunakhod mobile device appears to be of this type. For distant planets, where there are long transmission delays, such a device will be essential. As Cutler has suggested[34], as a subsystem, man leaves much to be desired. He proposes the production of remote observation means and manipulators that perform as well

108 Robot Mobility

as a man on a tether, in order to relieve astronauts of the necessity of working in space and on the other planets. The Russian programme would appear to be aimed in just such a direction.

Fully independent robots are required even for tasks much nearer home. Examples are operation on the reverse side of the moon, or investigation of the aftermath of nuclear explosions, above or below ground, or even inspection of drain and oil pipes from the inside.

Robots for vehicle driving[23-28, 38, 108-111]

Some of the possibilities and problems of vehicles driven by robots have been considered in Chapter 1. If a human is to be carried by the vehicle, safety is paramount.

It has been shown that a human car driver first looks as far along the road as possible, then back to the vehicle he is driving, then again along the road in front, and so on alternately. At a speed of 25 km/h a human driver finds it necessary to spend 60% of his time looking 50 m ahead, i.e. anticipating by about 7.5 s. He spends 80% of his time looking 30 m ahead, i.e. anticipating by about 5 s. It also seems that the human uses his central vision for obstacle detection and his peripheral vision as a guideline for general direction.

The possibilities of automatic traffic schemes have been considered, but up to the present there seems to have been more attention devoted to the problems of following the car in front, with steering and distance between vehicles automatically controlled[76-86, 104].

The Robotug system[1, 21, 83]

A form of automatic transport which has been widely used is the Robotug system. The robot vehicle is not completely free, since it is guided by signals produced by the field from a wire, buried some 2 cm below ground surface and carrying a current at a frequency of 2 kHz. However, the vehicle can be programmed to call at a number of destinations in turn.

The tug detects the position of the buried wire by two pick-up coils carried on the steering bogey on which the front, steering, wheel is mounted. The split-field d.c. servo motor used for steering is arranged to rotate the steering bogey until the two coils are symmetrically situated about the 2 kHz guiding wire. In this way, as the tug moves along, it follows the buried wire.

Other coils carried on the steering bogey can pick up signals instructing the tug to start and to stop, and also to signal the position of the tug in the system and to pass instructions to the controlling system, for example when it is necessary to choose the correct route so that the tug will reach the correct destination. A 400 Hz signal produced by an oscillator carried on the tug signals the position of the tug back to the central control system via buried coils which pick up the signal as the tug passes.

A counting system carried on the tug is used to determine the position of the tug with reference to a fixed point in the system. This system is used, for example, to determine the correct programmed stopping places and to determine alternative routes when the guidance cable diverges. Because a counting system is used, it is sometimes necessary to add dummy counts in order to equate the numbers obtained from alternative routes.

For up to 25 stopping places, the tug has been programmed by means of a number of toggle switches; but if more stopping points, up to 90, are required, then a patch board is plugged in, like that used with an analogue computer. The tug proceeds to the required stopping point via the shortest route, and it waits there until it is restarted on its way around the system.

A traction motor is used to propel the vehicle, and this is controlled by the track instructions. The driving speed is limited to between 3 and 10 km/h because of safety requirements. Electromagnetic solenoid operated brakes are automatically applied as soon as the vehicle is stopped, and an interlock system ensures that the brakes are disengaged before the motors restart.

The forms of vehicle said to have been use in this system include fork-lift trucks, large roofed vehicles and bi-directional vehicles. However, the usual form seen is a small towing vehicle which can thread its way through a crowded warehouse. The tug can carry a load of 1 t or it can pull two trailers carrying 2–10 t of materials, though longer trains have been seen.

The usual Robotug can haul 4.5 t up a 1 in 10 gradient, though a more powerful type is available which can haul 20 t on the level or 10 t up a 1 in 10 gradient. The drive motor is about 1.7 kW rating. A typical brief specification is as follows:

Maximum-draw-bar pull	454 kg
Sustained draw-bar pull	192 kg
Gross load (level surface and assuming rolling resistance of trailers is 23 kg per 1000 kg)	8500 kg
Unladen speed	3.22 km/h

110 Robot Mobility

Fully loaded speed	1.98 km/h
Weight (including 400 ah battery)	1000 kg
Brake (electromagnetic disc on motor armature shaft extension)	121 mm dia.
Battery voltage (lead–acid battery)	24 w
Battery capacity	400–658 Ah
Minimum turning radius	1.372 m
Towing hitch centre height	232 mm
Towing hitch pin diameter	25.4 mm
Rear tyres (solid rubber bonded)	305 × 108 mm
Front tyres (twin polyurethane)	203 × 44 mm
Over-all length	1.784 m
Over-all width	0.787 m
Platform length	1.245 m
Platform width	0.762 m
Wheelbase	0.838 m
Rear wheel tracking distance	0.610 m
Platform height	0.812 m
Over-all height	0.864 m

A system basically similar to the Robotug is now being used for the automatic carriage of goods in hospitals[66]. The carts weigh about 225 kg and they are 1.5 m in height. The guidance path wire is embedded in the floor, and it carries signals at either 6.5 kHz or 10 kHz. Snap-in boards fitted with diodes are used for guidance control, and a tiller is fitted so that a cart can be manually moved away from the guide track when necessary. The carts can move vertically by entering an elevator if necessary, and it is planned that they will automatically move to a charging area whenever the batteries need recharging. the track is divided into blocks, as with the Robotug, to prevent collisions, and a cart will restart its journey automatically after a stop of a predetermined length. It is hoped that the robot devices will be able to carry out the work of 60 people who at present would be required to transport the same quantity of goods around the hospital, which has a total of 800 beds. On this basis, the system will pay for itself in 7–8 years of service.

Robot urban transit vehicles[67, 119, 120]

The Bendix Company at Ann Arbor, U.S.A., has built a test track covering 6 ha, for investigations on automatically controlled, passenger-carrying, cars and vans. It is claimed that a substantial part of the ten billion dollars earmarked for public transport by the Federal Government over the next 12 years will go for unmanned, electronically controlled, cars.

At first, three small Swedish utility vans will be used in the Bendix experiments with electric battery powered motors replacing the original engines. The present test track consists of a 100 m diameter guideway, joined by several feeder lanes, giving a total track length of about 1.5 km. The track has three passenger stations.

The control system is designed to maintain a spacing of 10 s between vehicles travelling at a maximum of 50 km/h, though it is hoped later to reduce the spacing to 5 s. The buried guidance wires are spaced 60 cm apart under the roadway, and they are transposed every 7.5 m, or every 4 m on the sharper bends where the speed limit is lowered to 25 km/h. At each of the crossovers the magnetic field, produced by the 145 kHz current flowing in the wires, falls to zero; these zero points are detected by the vehicle control system and the information is transmitted back to the central computer control by means of a frequency-shifted 96 kHz carrier signal.

Wayside communicator stations are situated along the tracks to transmit and receive the control signals. In case the main guidance track system fails, an independent back-up system using magnetometers for vehicle position sensing is also fitted. The magnetometers are spaced by 30 m in the fast zones and by 2.5 m in the very slow zones. If the spacing between any vehicles falls below one of the magnetometer block distances, then the vehicles are automatically braked to standstill.

The commands to the vehicle from the central computer are 'Start', 'Brake', 'Proceed to station X', 'Run at constant speed' (between 25% and 110% of the nominal maximum speed), 'Switch lights on (or off)', 'Switch air conditioning on (or off)', 'Switch windscreen wipers on (or off)'. The messages take the form of a 16 bit word followed by its complement and include three 'identification bits'. Each of the vehicles can receive up to four commands each second, at a rate of 4800 bits/s. The central command computer signals a destination to the vehicle, and the destination identification is stored by the vehicle, though it is intended that in the future it will be possible for the passenger to enter his own required destination by means of push-buttons inside the vehicle.

The vehicle guidance is by means of two coils mounted on the bumpers, each about 40 cm from the centre line, which pick up current induced from a third 'guidance' wire buried in the roadway, rather like the Robotug system. To provide for route selection, different sections of guidance cable operate at different frequencies, and there is some overlap of the cable sections to ensure continuity of guidance. Continuity of command signals is ensured by the provision of two pairs of receiving and transmitting coils on the vehicle, one pair at the front, one pair at the rear.

A special tone is transmitted to the vehicle to signal it to slow down with a controlled deceleration pattern and to stop. The controlled deceleration pattern is at present carried in a diode matrix, though it is planned to use an integrated Read-Only Memory for this. The deceleration is controlled by comparison of the signal from a distance measuring device, which gives a pulse every 23 cm of travel, with the command signal stored in the memory, the difference being used to control the braking and propulsion on the vehicle in order to stop within 15 cm of the required point.

A tachometer is used for speed measurement at speeds below 8 km/h. Above this, the intervals between the guidance wire cross over points are measured. Estimates of the cost vary between 5000 and 8000 dollars for each vehicle.

The central computer used in the control system is a type PDP11/20, with 12 000 words of storage, which can later be expanded to 32 000 words if necessary.

Safety with the Robotug

The safety measures which have been taken with Robotug installations are of interest for their applications to the more general mobile robot in addition to their importance in the present practical Robotug applications.

Some of the safety precautions which have been taken are specific to a track-following vehicle. These include arrangements to stop the vehicle in the event of failure of the track guidance signal current or if the tug steers away from the track and again loses signal. However, other of the safety precautions are of interest in the wider application of the fully mobile robot vehicle.

For example, a light bumper is fitted to the truck, projecting some distance in front. Deflection of this bumper stops the tug before it can reach the obstruction. If, however, the first bumper fails, there is a second, more robust, back-up bumper which again will stop the tug.

If the voltage level of the supply battery carried on the tug falls too low, then naturally the control system will become unreliable. Consequently, arrangements are made to stop the tug in the event of the battery becoming discharged below a safe operating level. In a mobile system overload is quite a possibility in unforeseen circumstances, and it is necessary to provide overload trips and fuses to avoid the possibility of danger. One obvious danger which could be encountered in warehouse applications is that of fire in the event of a severe overload, and so such precautions are clearly very necessary.

In order to warn humans of the approach of a tug, flashing lights or klaxon horns, either mounted on the tug itself or at points along the track system, are used. The tug can operate warning signals along the guidance track route automatically by its approach. It can also operate traffic lights or stop other traffic—for example, at an intersection with a road used by normal vehicles—and it can automatically cause doors to open to permit its passage. One advantage of the latter facility is the heat conservation which can be achieved by avoiding permanently open doors.

The guidance track is divided into blocks, as on the railways, so that a tug cannot enter any section of the route unless the preceding vehicle has already cleared that section. In this way one tug is prevented from running into the rear of another tug or train. This precaution is also extended so that a tug cannot enter a track section unless the next-but-one section ahead is also clear.

The main track guidance system has a memory device which automatically monitors the position of every truck on the system and displays the information on a mimic diagram if required. The memory system is also used to release a tug waiting at the start of a section as soon as the track is clear.

Other typical reasons for halting a tug are that it is on a converging section and there is a tug already on the way to the convergence on the other track, or that there is a crossing ahead and there is a tug heading for the crossing on the other track. Each block must clearly be longer than the train length. It is important that the number of blocks around any closed loop on the system be odd, since otherwise there is a danger that a complete seizure can occur when every alternate block around a loop is occupied, and no further movement will then be possible.

The advantages claimed for the Rotobug system are (a) expensive manpower is conserved; and (b) because of the elimination of the fickle human element, there are fewer accidents— for example, those caused by attempts to cut corners with a long train of trailers.

Control is very close and the time taken for any journey is accurately known. This feature is most important, as is realised by anyone who, like the writer, has made attempts to apply closed-loop servo techniques to production control and organisation. The attempts always fail because of the extreme time variation of the human being when called upon to carry out any task involving the transport or transmission of goods (or of information) from one place to another.

Not only is it possible to cater for the safety of humans with such a robot system, but it is also possible to reduce the accidental losses of goods caused, for example, by driving over weak floors or over the edge of a loading platform.

Similar steering systems to those of the Robotug have more recently been used by the Road Research Laboratory to control a Citroen DS 19 at 130 km/h by means of an electro-hydraulic steering system and also to control Mini and Cortina cars by means of d.c. motor steering[28].

Automatic farm tractors[18, 17, 47, 63, 121]

It has been pointed out by Brooke[43] that the process of steering a farm tractor is the most time-consuming, yet in many ways the least important, of the tasks of the tractor driver. Steering the tractor can interfere with the driver's correct actions in, for example, ensuring the correct distribution of seeds, fertilisers or pesticides, or in the optimisation of a threshing process. In other cases such as harrowing a human driver is wasting his skill on the performance of a repetitive task.

There have therefore been a number of investigations into the possibility of eliminating the operator completely or at least easing his task to the point where he can supervise several machines at once. Warner[45] has pointed out that automatic operations might make it possible to work in adverse conditions of weather and light, and make it possible to complete work speedily while the optimum conditions of soil, weather and crop exist, even though these are only short-lived.

In some farming applications automatically controlled tractors can not only relieve the human operator but also produce an improved result. Examples are drain laying[41], beet harvesting and rowcrop thinning[42], though further work is required before the combine harvester can be controlled efficiently.

A good accuracy of steering is required, of the order of a few centimetres though, in general, this needs only to be with respect to the adjacent pass[43]. Buried guidance wires have been used, and these must be spaced by no more than 6–12 m.

If, however, the ratio of the vertical field strengths on each side of the tractor can be determined, the cable spacings of 30–60 m can be used. Three ferrite cores with vertical axes have been used to detect the field at 1 kHz from cables carrying a current of less than 100 mA.

There has been some investigation of a method of guidance in which a furrow is detected by an ultrasonic method and then followed. However, while this method is suitable for use on a hard surface such as tarmac, it does not work too well on a rough ploughed surface. The use of optical methods has been suggested, and optical methods have been used for drain-laying control. There has been some work on the use of Doppler methods for radar measurement of tractor speed, and these devices have been suggested as an alternative to capacitive proximity switches for obstacle detection. It has been found that a tractor running light on grass at 11 km/h will stop in about 1.25 m with the rear wheels locked and the drive disconnected.

The steering system which has been used on automatic farm tractors is hydraulic[44]. It has been suggested that, for each driver displaced, a sum of between five and ten times the driver's yearly pay becomes available for expenditure on the capital cost of the automatic guidance equipment; and that, since the U.K. output of tractors is about a quarter of a million annually, there should be a big market for a successful robot system[45].

If a buried cable is required for guidance, the cost is of the order of £25–40/ha[44]. A realistic price for a control system including only the basic essentials has been estimated at £400–500. It has been suggested that for the present it is best to have the control in the form of an add-on unit for a standard tractor and that, later, special automatically controlled tractors will be required.

It becomes clear that, like most other robot applications, farming by use of robot tractors is being held back not by lack of techniques but by lack of expenditure on development. While the use of underground guidance cables is probably a satisfactory approach for use on smaller farms, it is probable that alternatives such as optical methods will find a more extensive application. Eventually we can expect fully automatic robotic machines to be used on the farm, not only for field work but also for care of livestock. Already there is investigation into the problems associated with livestock environment control[46], and such work is likely to spread.

Flying robots

The traditional robot of fiction walks, but it does not usually fly. Practical robot devices are more versatile. Unfortunately. For the best-known example of a completely mobile, completely independent robot is the guided weapon, the robot flying missile[39, 40]. These devices seek out their target under difficult conditions and they seek it out much more unerringly than could any human.

In space flying robots spy on activities on earth. Here, however, they also carry out much more peaceful tasks as the relaying of television programmes or the exploration of the moon. The robot is much more versatile than is the human in such respects. This versatility is likely to increase as we discover more about the production of more complex robot nervous systems. Already it is possible to construct automatic aircraft pilots[87] which can not only control the aeroplane in level flight but also take off and land automatically.

There is now the additional possibility of the mobile robot based on the principle of the hovercraft. The principle has been used on domestic appliances and on lawn mowers, but not yet for the suspension of mobile robots[62]. A widely used form of flying robot is the balloon-borne radio- and radar-sonde system[88-90] used to signal back upper atmosphere data for weather forecasting, though its direction of travel is not at all self-determined, being controlled by the wind direction.

Radio-controlled pilotless aircraft have long been used for purposes such as target practice, where it was obviously not possible to use a human pilot. For example, the Queen Bee target aircraft[91] used in the early 1940s was simply a modified Tiger Moth conventional piloted biplane. The Queen Bee was controlled from the ground by means of a set of ten buttons, or sometimes by a telephone-type dial. It was found possible to land the Queen Bee, fitted with floats rather than with wheels, quite successfully by remote control even if the sea was quite choppy. An improved version of this plane was known as the Queen Wasp, and similar control systems were used on the radio-controlled target speed-boats, the Queen Ducks and the Queen Gulls. The latter led to the more complex Restless system, in which radio-controlled speed-boats moored at sea could be started and controlled from the shore to attack warships.

A target aircraft developed at a later date was the Jindivik, largely made and tested in Australia and a product of the early 1950s[92]. This aircraft could take off from a gyroscopically controlled, recoverable trolley. The pneumatic actuation was powered by air stored at a pressure of some 14 000 kN/m^2, with filtering and pressure reduction to about 4000 kN/m^2. Electrical power for the Jindivik came from a 6 kW d.c. shunt generator across which was floated a 12-cell lead-acid accumulator. A carbon-pile regulator controlled the voltage to 27 ± 0.5 V. The main power supply came from a gas turbine engine. There have been many more recent developments, including proposals for the use of such pilotless aircraft in combat[93, 94, 96].

Space robots[68-70, 97, 98, 117, 122]

In the absence of space shuttle systems which can ferry repairmen to satellites in orbit, there have been design studies on expendable robots, or remote space manipulators for extravehicule tasks. The possibility of remotely controlled space robots was realised early and, indeed, the Surveyor 3 unmanned moon flight was equipped with a digger controlled from earth. It was found possible to scoop up moon dust samples and to deposit them within 6 mm of the required position. However, the potential value of such remote manipulation was demonstrated dramatically in January 1968 on Surveyor 7, when the digger was used to correct an equipment failure which had occurred unexpectedly on the moon.

It has been found at the Argonne National Laboratories that an operator in shirt-sleeves with a master–slave manipulator is as dexterous as is a space-suited operator working directly, both taking three times as long to perform a task as they would have done directly by hand. Remote manipulators have been proposed for any space applications where there is a hazard to humans, or where endurance is required, or where there is an advantage of cost, weight or probability of success. Such remote manipulators have been called Androidal Teleoperators, or Android for short, but one hopes that this will not be generally adopted, since the term 'Android' has a very special and definite meaning.

The form of space manipulator envisaged at present has seven motions: one for grasping, three for translatory motion and three for angular motions.

The Surveyor system has four motions, each step-controlled from earth, the only feedback to the operator taking the form of still pictures which take about 1 min to process. The control is thus very slow. In general, bilateral manipulators, i.e. those having feedback to the operator, involve task times of some 3–10 times that taken by a direct-hand operator, while unilateral manipulators without feedback take some 30–100 times as long. The penalty paid for the feedback is a weight of some 45 kg.

Studies have been made of preliminary design of a standardised general-purpose electrical space manipulator for use on both manned and unmanned missions. Typically, the vehicle must be able to dock to a satellite to transfer cargo, to open hatches, to replace satellite electronic modules and to release the satellite after testing the system. It must be able to repeat the mission ten times in 2 years. It must be able to maintain maximum grip for 30 s without exceeding a temperature of 100°C. The control transmission time delay will be between 0.24 and 1.0 s. Studies indicate that the requirements can be met.

A design published in late 1969 had two arms, one one each side of a T.V. camera, the total weight of the vehicle including fuel being almost 450 kg, while the normal operating power and peak power were 200 W and 1000 W, respectively. A close-up camera on a semi-rigid tether would also be provided. Studies such as this bring closer

Robot Mobility

the day when we shall be able to send true robots into space, reporting back to us but not under direct movement-to-movement control. The control delays involved make it quite certain that this type of semi-independent robot, obeying general orders rather than detailed control, will be required.

Where there is a delay between a human operator and the device being controlled, as when the device is in space or on another planet, it has been suggested[68] that a rough estimate of the time required for completion of a task is given by

$$T_c = 1.6 T_0 + \frac{T_d}{1.65}$$

where T_c is the required time; T_0 is the time required if there is no delay; and T_d is the delay time involved. Consequently, for ground control of a manipulator in orbit around the earth, with a 0.75 s delay involved, the task time would be doubled.

It appears that in future work the 'operator' will specify sub-goals for the space robot, but it will not be practicable for him to be in direct control[71-74]. Sheridan and Ferrell have investigated the problem by simulation using an AMF model 8 master–slave manipulator driven by stepping motors via an augmented PDP8 computer. The human specifies sub-goals and procedural restraints, and the robot machine does the rest automatically, so overcoming the problems introduced by transmission delays.

A little-publicised significant robotics feat was that of the Russian Luna 20, which landed on the moon, drilled into the rock and lifted rock samples into a spherical capsule, which then returned to earth and landed in a blizzard in Kazakhstan, all completely unmanned by humans[95, 123].

Mobile manipulators[99, 100, 102]

A manipulator can be thought of to some extent as merely a modern form of versatile crane. Certainly some forms of shovel and of crane are capable of a wide variety of actions, and the line between these earlier devices and the more recent manipulators is very blurred.

As an example, the GE Man-Mate is in effect a crane which can be fitted with various forms of gripping device, such as mechanical gripping jaws or vacuum cups for lifting glass. The advance in such machines is really an advance in fineness of control, since the operator has a delicate touch and there is a feedback from the actuator to the operator. The advantage claimed for such machines is mainly reduced operator fatigue in handling heavy tasks.

The advantage of making such a manipulator fully mobile is obvious, provided that the main obstacle of increased cost can be justified. One application where the cost can indeed be justified is to investigation and repair following any accidents in a nuclear reactor or similar system. A device of this type which was used for a time at the Nuclear Rocket Development Station in Nevada was called the Beetle. The Beetle carried a crew of two in a massively shielded cab, and it had two arms, each about 5 m long. This device proved to be too costly to operate, and later devices have in general, used an operator, or operators, who can be situated remotely in a safe environment.

A manipulator of this later type was the Mobile Remote Manipulative Unit or MRMU. This weighed only about 11 000 kg, compared with the 80 000 kg of the Beetle. It was controlled by radio and a number of television cameras provided visual feedback. The basic vehicle used for the MRMU was a converted army cargo carrier tracked vehicle, which could be driven at speeds of 50 km/h, about four times faster than the Beetle, since the heavy shielding of the Beetle was no longer required.

A form of free-ranging vehicle used for routine transportation in radioactive environments is the GM/PaR Little Ranger. This has a platform 1 m square which carries a single arm and two cameras in a stereo arrangement, on a central column. The platform is mounted on a tracked vehicle and is connected to the human control station by a cable, which has the effect of limiting the permissible movement of the vehicle.

The Hughes Mobile Robot, or Mobot[5], carries its own battery power supplies, and so a greater mobility is obtained, though a trailing cable is still used to carry command signals. The two arms each have three joints, and each of these joints is capable of a full hemispheric motion. There is consequently an extreme of flexibility. Other features which can be added are booms for the cameras, a jib crane and a fork-lift. The Mobot is mounted on three wheels, the single rear wheel being used for steering. The maximum speed is 5 km/h, pneumatic brakes being automatically actuated when the Mobot stops. The speed of travel and of the arm motions is controlled by foot throttles, while toggle switches provide the main operator's controls. Commands are transmitted over a three-conductor cable, synchronous commutating switches being used for multiplexing.

Some approximate details of the Mobot are as follows:

Maximum load   12 kg in any position of the arm
Pressure of finger pad 0–100 kN/m^2 (adjustable)

Robot Mobility

Finger area	35 cm²
Maximum finger force	350 N (full finger area engaged)
Speed of finger closure	8.5 cm/s
HAND Rotate stall torque	45 Nm
Rotate speed (no load)	10 rev/min (continuous)
Telescope stall thrust	200 N
Telescope speed (no load)	12 cm/s
Stroke	9 cm
WRIST Torque	55 Nm
Speed	2 rev/min max. (hemispherical coverage)
ELBOW Torque	165 Nm
Speed	0.75 rev/min max. (hemispherical coverage)
SHOULDER Torque	330 Nm
Speed	0.33 rev/min max. (hemispherical coverage)
Arm clearance	16 cm diameter hole

The Self-Propelled Anthropomorphic Manipulator or Sam[103] which was built at low cost, is a four-wheeled, remotely controlled vehicle carrying an articulated arm mounted on a semicircular track. The manipulator can operate from ground level to nearly 2 m above the ground. The manipulator arms are Rancho Los Amigos orthotics and they are remotely controlled by an exoskeleton worn by the human operator. The operator wears a head harness which controls a T.V. camera carried on the manipulator. Command and control is carried out via a commercial 64-channel PCM-FM radio link.

The Mascot mobile manipulator[7-11]

Some of the most outstanding results with mobile manipulators have been produced since 1959 by the Servomechanism Laboratory of the National Committee for Nuclear Energy (CNEN) at Casaccia Italy, under the direction of Carlo Mancini. This work will be described in some detail here, since it provides an excellent model for future work on independent robots. Based on the earlier work of the Remote Control Division of CNEN, the research culminated in the production of the Mascot mobile manipulator, which is now made commercially by Selenia of Rome. The name Mascot is short for MAnipolatore Servo Controllato Transistorizzato.

From the start Mascot was designed on an engineering basis. It was estimated that the cost of covering a working volume with a mechanical form of manipulator was about 1000 dollars/cm³. Since the cost of any mobile robot type of manipulator was initially estimated at about 50 000 dollars, it would have to be capable of covering a minimum working volume of over 50 m³. It would also have to be capable of handling a work load of 23 kg and of lifting the load to a height of 4 m, and have a dexterity comparable with that obtained with a mechanical manipulator.

The mobile slave equipment was mounted on a remotely controlled trolley powered by electro-hydraulic actuators, and it was also to be provided with a stereo T.V. camera. Each of the two arms has seven possible motions: three in translation and three in rotation and the seventh is the squeeze motion.

A notable engineering feature of Mascot is that identical modular servo controls have been used for all of the seven movements of each arm. Stabilisation of the servos is by feedback of a velocity signal. It is of interest here to give the specification of the standardised servo system:

Maximum torque	20 Nm
Maximum speed	70 rev/min
Maximum self-synchronising angle	2.5 rev
Starting friction torque	0.1 Nm
Compliance	0.002 rad/Nm
Bandwidth	25 Hz
Full load damping factor	>1

The two-phase motor used in the servo system was specially designed and built. The specification is as follows:

Supply frequency	50 Hz
Stall power per phase	180 W
Maximum torque	0.5 Nm
No load speed	2900 rev/min
Stall acceleration	80 000 rad s^{-1} s^{-1}
Friction torque	0.001 Nm

Both the motor and the controller were designed for continuous duty with maximum applied torque, an air flow of 30 m³/min being required for cooling the motor in an ambient temperature of 25°C. The fan was arranged to cool the motor whenever a fixed value of temperature is obtained. The average temperature of the motor is minimised by controlling both fields simultaneously, rather than just one, and this helps to increase the motor life. The maximum power output from each of the channels is 500 W.

Great care was taken with the construction and with the materials used in the early Mascot. For example, stainless steel control cables, corrosion-resistant aluminium and radiation-resistant greases were used. The tension of the operating is maintained, despite movements of the relative positions of the lower and the upper sections of an arm, by the use of special cams.

The feature of failure to safety was included, all motions being automatically locked by switching off an electromagnetic brake release and so allowing a spring to operate the brake in the event of failure. A failure is indicated either by an increase of the main synchro control voltage above a preset maximum value or by a decrease of this voltage to zero.

The weight of the slave unit is 775 kg, the over-all size being 125 cm × 110 cm × 170 cm. The maximum speed at which the tongs can move is 75 cm/s, with a starting friction of 0.25 kg and a compliance of 2.8 cm/kg.

As a result of this work, it was suggested that the speed and sensitivity of the movement should be increased, that the servo friction should be reduced, and that the steel operating cables should be replaced by tapes.

In the early work the weights of the arms were balanced by counterweights, and this method had the disadvantage of increasing the inertia of the system. In later work the weight was balanced electrically with torques generated by the servo driving units. The new approach also gave greater freedom in the design of the arms. The friction and the inertia of the arms was largely due to the servo drive units, and these factors were also reduced by the use of feedback torque signals. Such methods have improved the feel of the system to the operator, under both dynamic and static conditions.

At the same time, the required number of leads between the master control unit and the slave unit was reduced. This gave a greater flexibility of movement to the slave unit. In addition, the dimensions of the servo drives, and therefore of the arms themselves, were decreased. This had the incidental advantage of improving the versatility of the system, since a wider range of tasks could be undertaken with a smaller unit.

The power amplifiers which have been used on the later devices are of the switched high-efficiency type, with natural cooling. One object of the adoption of the new forms of power amplifier is a reduction of the size of the control power amplifiers to the point where they can be installed directly on the body of the robot rather than remotely. Since only the low-power control signals then have to be taken to the robot unit, smaller and lighter trailing cables can be used, and eventually it is hoped that the complete flexibility of radio control of an independent robot can be used.

Later Mascot work[12, 13]

No attempt will be made here to present a comprehensive treatment of the methods used on Mascot. A very thorough treatment is given by the publications of the Italian workers, who have clearly taken pains to make sure that the information is available to all.

On the later work on Mascot the two-phase servo-motor was used for the following reasons:

1. The motor is rugged and requires little or no maintenance.
2. The ratio of maximum output torque to friction torque is high. This leads to a high system sensitivity.
3. The drive is smooth, since the torque is constant over the whole revolution of the motor shaft.

Instead of a single larger motor for the drive, four coupled motors are used. This leads to a compact arrangement with a reduced overall moment of inertia and an increased surface area for heat dissipation, permitting a high value of overload capacity. The motor supply is 115 V, 60 Hz, and the other details of the standard motor are:

Maximum control phase power	33 W
Starting torque	0.1 Nm
No-load speed	3500 rev/min
Starting acceleration	34 500 rad^{-1} $^{-1}$
Power output	10 W

The maximum control winding voltage is 150 V, while the reference winding is normally operated at 100 V, being switched up to 150 V when maximum torque is required. The maximum power dissipated is 70 W at full voltage, falling to 25 W at rest. The slave servo drive unit output torque is a maximum of 20 Nm, and it can be maintained for 15 min at an ambient temperature of 20°C.

Position measurement is by means of a variable phase signal from a 400 Hz linear synchro, the range being $\pm 60°$. The synchro is supplied with a frequency of 2 kHz, the same supply being used for the two-phase tachometer generators used for speed signals. Compared with the earlier Mascot arrangement, the total number of leads between master and slave is reduced from 55 to 33 by the use of the new arrangement. Tests with film potentiometers for position measurement were unsuccessful.

The drive motors are built directly on to the drive gear-box, a reduction ratio of 38:1 being used between the motors and the output shaft, while a ratio of 10:1 is used between the output shaft and the linear synchro. The starting friction torque is 0.03 Nm and the outside diameter of the gear-box is rather over 12 cm.

The servo amplifier details will not be presented here, as they have been given elsewhere. However, the time-division form of power amplifier which was used should be specially mentioned, since this gives a very small and simple arrangement which has low losses and can be cooled by natural convection. The advantages of this system appear promising for eventual use in a power amplifier mounted on the mobile device, controlled by time-division pulses over a radio link. At present up to 300 m of cable can be used.

The research work on the Mascot device has led to its commercial production by Selenia of Rome. In the commercial model it is stressed that the robot is virtually self-repairing, since one of the manipulator arms can be repaied by the other. There are two separate failure-to-safety features. In the event of power failure, the arms and hands are locked in the last position. If there is a minor failure— for example, a failure of the cooling system—the operator is advised 90 s before an automatic shut-down, so that a task can be completed or avoiding action taken. To reduce operator tiring, the feedback ratios used are adjustable, and if required the operator can lock the control loop of one arm by use of a foot control, while he concentrates on the action of the other arm.

The research work leading to the successful production of the commercial Mascot robot manipulator is certainly deserving of great praise, not only for the outstanding engineering results achieved but also for the unselfish way in which the results have been published freely and the way in which the difficulties encountered, as well as the successes, have been openly discussed.

Perhaps the best praise that can be given for the final result is to quote Ballinger[6], who stated that the machine creates a humanistic impression which generates an impulse to speak orders to the machine rather than to the operator.

If future developments go as expected, it will almost certainly be possible to do just that.

Near field control[49]

Where a robot device must be controlled by a human, in some cases the robot has been linked to the control console by wires. This has the effect of limiting the mobility of the robot and of the operator. It also calls for care in order to avoid the tangling of the control cable. In some cases it has been necessary to arrange for the control cable to drop from overhead to the robot.

Greater mobility of both robot and operator is possible if a radio link is used for control. A typical example with which the writer was associated some years ago was the control of a ropeways trolley on a long track stretching out to sea for oil drilling, the frequency of radio control being of the order of 60 MHz.

Where a robot device must be controlled or communicated with in space, there is, of course, no alternative to the use of radio communication. The problem of delay in the control achieved by this method has been mentioned elsewhere.

There is a large range of robot control and communications problems where the distances involved are only of the order of 100 m at the maximum, but a complete flexibility of movement of the robot and the human (or of a second robot) is desirable. Radio communication introduces problems of finding spectrum space and of avoiding interference. Communication by light waves is not always easy because of the severe directionality. Ultrasonic communication is a possibility, but it does not seem to have been widely exploited except for some applications to crane remote control.

One system ('Telemotive') which overcomes many of the problems in control and communication over distances up to 100 m is the use of near field induction effects at medium frequencies of a few hundred kilohertz. An induction field falls rapidly in strength as the distance from the source antenna increases, being approximately inversely proportional to the cube of the distance from the source. The effect can be detected at distances up to about one-tenth of a wavelength from the source. Consequently, an induction field system is fundamentally ideal for use in the type of application envisaged. Frequencies of 250–400 kHz have been used, the transmitted power being 60 mW and the sensitivity of the receiver being about 0.05 m.V. A range of up to 60 m is obtained with such a system, and repeaters can be used if this has to be extended.

Such a system has been found to give immunity from radio-frequency interference while meeting with Post Office licensing approval. It is possible to design for failure to safety and easy and immediate maintenance by untrained personnel even though the belt-mounted operator transmitter is small and light in weight. The control channels have been spaced by 125 Hz within a total bandwidth of 4 kHz, and one user is said to have 20 systems in use in the same shop, using different carrier frequencies and separation by distance to ensure absence of mutual interference, and in some cases two separate operators can control on separate frequencies on a 'first come, first served' basis. The near field system is now being widely used in a variety of applications.

The Rivet machines, [14-16, 101]

The term 'telechiric' has been used to some extent for remotely controlled devices. It is derived from the Greek for 'distant hand'. A machine designed at Harwell was known as Rivet, from Remote Inspection Vehicle, Telechiric. This device has roughly the profile of a crawling man while it is moving along, but an antenna fitted with a T.V. camera can be erected when required. An arm having a reach of about 150 cm radius can lift loads of 35 kg. While moving, it can climb over obstacles having a single step height which is half as long as the vehicle track—that is, four times the height which can be cleared by a tracked vehicle. The vehicle is very manoeuvrable, being capable of climbing stairs at an angle of 45° and of turning in a corridor 1.2 m wide. According to Ballinger[6], Rivet can enter an office, pass through the knee-hole of a desk and then climb onto the top of the desk.

Thring has demonstrated various versatile crawling machines, some of which can climb stairs.

Mobile machines controlled by a large computer[2, 3, 19, 22, 37]

In various universities around the world it has been possible for departments to obtain substantial financial support which has enabled them to carry out work on the control of a fully mobile machine by means of a large digital computer.

Examples of such investigations are the work at MIT[33] under Minsky and Papert, the work at Stanford University under McCarthy[20, 35, 36] and the work at Edinburgh under Michie[4, 32, 65, 107, 124].

At MIT a quite complex jointed artificial arm and hand[33] has been controlled by a PDP6 computer receiving information from a television camera. The hand can grasp and pick up blocks of various sizes and assemble these into a definite defined shape within a given area. The shapes can be collected in a specified sequence if required.

It is difficult with such complex assemblies to compromise between selecting an over-simple task and selecting an over-complex task. At Stanford a fully mobile machine, mounted on a steerable trolley, has been used. There has been no attempt in the early work to build the computer into the mobile arrangement, and a heavy cable is taken overhead to the computer. The trolley is driven by two separate step motors, one for each of the driving wheels mounted on each side. The TV camera mounted on the trolley can also be moved by a step motor. In early work no hand was fitted at Stanford but the trolley could push objects along. An optical rangefinder was fitted, to supply further information to the computer, via the cables. It was proposed to fit radio links eventually to replace the cable links. A large time-sharing computer was used.

Edge detection in the electronic system following the TV camera and a process of differentiation, segmentation and reconstruction of the segments into lines are used in the computer to construct an equivalent line diagram for the object viewed. Navigation was by dead reckoning from the starting position, though this leads to errors caused by slippage and by uneven ground.

Recent work at Edinburgh University on the FREDDY device in the department of Machine Intelligence and Perception has used a Vidicon TV camera for vision. A large two-fingered hand is controlled from a computer in response to the TV signals. The work is concerned basically with the programming of a computer to achieve an integrated cognitive system—that is, a system which can construct and manipulate abstract models of the external world[124, 127].

The Hitachi HIVIP

The Japanese Hitachi robot arm, HIVIP Mk 1, is an assembly device which views an orthographic projection of an assembly which it is to produce, views the component parts from which the assembly is to be made, and then establishes an assembly sequence and directs a seven-mode manipulator in its assembly process. The Hitachi work has been described by Ejiri et al.[50, 105, 106]. As opposed to the earlier industrial robots, which required a detailed program of the precise movements to be prepared manually in the correct sequence, the instructions to HIVIP merely take the form of a three-view drawing of the required assembly, similar to the three-view drawing which would be given to a human who had to carry out the assembly task.

The robot is composed of three sub-systems: eye, brain and hand. The assembly drawing is viewed by one Vidicon TV camera, while a second TV camera views and locates the collection of parts awaiting assembly and arbitrarily placed on a table.

Each TV picture area is divided up into 76 800 elements, with 240 vertical and 320 horizontal, while the amount of light reflected from each of these elements is encoded in an analogue to digital convertor using a 5 bit, 32 level binary code.

The required control movements are computed by a HITAC 7250 computer, having a 32 768 word, 16 bit core memory, a 2 μs cycle time and a 4.5 μs add time. The computer blocks off the scene viewed by the camera lens until only the single required image is being viewed. The remaining picture information is then thinned in three steps: first four picture elements at once, then two at a time, then individual elements being fed to the computer. The computer itself selects a suitable level so that a two-level, black and white picture is obtained, and it can also differentiate in order to obtain the outline of objects.

The seven degrees of freedom of the articulated arm and the parallel gripping jaws of the HIVIP robot are under the simultaneous control of seven independent digital servo systems, each using synchro-resolver position detection and thyristor-controlled d.c. servo motors.

The assembly instructions to the robot are supplied in the form of a three-projection drawing, which is scanned by a TV camera and used to derive a list, within the computer, of nodes, branches and loops. From the list the three-dimensional configuration of the assemblage is determined, the usual problems caused by hidden lines being eliminated by the computer program. The nature of the structure is then computed by the program and the structure is broken down into the sub-sets of component parts required for the assembly.

The loose component parts on the table are detected by scanning with a Vidicon TV camera, and the information is supplied to the computer, coded into a five-bit form. Colour is not used at present. The position of each of the objects on the table is determined and memorised by the computer; the parts necessary for the required assembly are determined; the necessary order of assembly is computed as the particular reverse of the disintegration process which would require the minimum number of assembly movements; and two parallel opposing surfaces on the object, suitable for gripping by the robot fingers, are decided upon. The necessary paths and movements of the hand and the angles of each of the joints are then computed, and the seven movements are then simultaneously and independently controlled by the computer to give a continuous-path movement.

The software of the robot exceeds 400 000 words of memory, and it is stored in the magnetic drum memory of 512 000 word capacity, the Fortran language being used. However, as long as 240 s can be taken for image processing, 20 s for drawing recognition 50 s for the recognition of each object, 10 s for decision making, and 180 s for movement for an average assembly. It has been suggested that new and faster processors are required, and that perhaps these will make use of parallel operation.

The Stanford project[53, 112-115, 126]

In the Stanford University Artificial Intelligence Project a mechanical arm was used to stack cubes which had been located by an edge tracing program. In the early work[51, 54, 52], it was only possible to detect and to outline well-illuminated white cubes on a black table, but later work was devoted to detection in a less carefully controlled environment.

A Vidicon scanner tube gives a set of 333 × 256 samples of image illumination intensity, 60 times each second. The intensity of each of the samples is encoded as a 4 bit, 16 level number, the effective width of the intensity 'window' being adjustable within 1 ms in order to obtain optimum results by obtaining the best compromise between the dynamic range and the resolution obtained. For maximum resolution, the 16 levels of quantisation cover only a $\frac{1}{8}$ V portion of the full 1 V working range of the video amplifier, while for maximum dynamic range the 16 levels occupy the full 1 V range. In this way the effective resolution obtained is increased to the equivalent of 128 levels, 7 bits. The information is fed into the high-speed data channel of the computer used, which has a capacity of 24 million bits/s.

The PDP6 computer controls the pan and the tilt of the camera; the lens used and, hence, the focal length, by rotation of the lens turret; the colour filter used; the focus; and the tube target voltage. The lens iris is at present set manually. The colour filter wheel is fitted with three colour filters and one neutral density filter, selectable in 0.2 s, and it is mounted between the lens turret and the Vidicon tube.

The camera target voltage used is selected by the computer in 64 steps, between 0 and 50 V, though some 10 TV frames are required for image stabilisation after a change of the target voltage. The voltage is not allowed to exceed that which will produce a given safe value of average signal current.

In the early work at Stanford it was necessary to set the camera up manually so as to emphasise those features needed in a particular task, and the edge follower procedure could only handle simply closed contours, while a simple 3 × 3 gradient operator was used for edge detection. The more recent work has used a regional operator, covering between 32 and 177 raster points. as described by

Hueckel[55]. This can detect weak edges over a considerable area, despite noise on the picture. All edges of an object need not be traced at the same time, since the program used makes provision for the later merging of end points of lines whenever a previously encountered point is reached. Gradually, by means of various safety procedures built into the program, a completed picture of the edges of the objects being scanned is built up in the computer memory.

In effect, the procedure as described by Pingle and Tenenbaum[56] has the advantage over some of the earlier methods that, by means of feedback, the scanning is adapted to the deficiencies of the scene being observed so that the optimum results can be obtained. Falk has described methods of obtaining information from imperfect line drawings of assemblies[57, 125].

A number of papers have described the sort of programs, languages and time-sharing used at Stanford[51-54, 58].

The arm used in some of the work at Stanford has been a prosthetic arm produced by the Rancho Los Amigos hospital near Los Angeles[59, 64]. The six joints are each provided with a potentiometer for position feedback to the control system, pulse-width control of the speed of movement being used. A two-finger parallel grip hand has been fitted. A PDP6 computer and a PDP10 computer have been linked together for control, sharing 128 000 bits of core memory, and with time-sharing on the PDP10. A special form of ALGOL language, known as SAIL, has been used. It has been demonstrated that this arrangement can stack four multi-coloured cubes in such a way that each side of the stack has four different colours.

The Stanford robot has been programmed so that it could position a ramp, up which it could then drive in order to collect a box positioned on a raised platform[75]. The robot takes some 30 min to complete the task, 20 min being required for computation on a time-shared SDS940 computer and 10 min being required for the necessary physical movements. It is perhaps important to note that it took about 1 year of detailed preparatory work before the computer could carry out such a task.

The recent developments at Stanford are most impressive, but it must be remembered that a great deal of computer capacity is necessary. Nevertheless, work such as this can give a guide to future work, not only on the simplification of procedures and programs and on improved visual detection methods, but also on possible approaches to the use of parallel processing of information in order to speed up the operation and to make it possible to carry out a number of the operations simultaneously.

Legs, wheels or tracks

The wheel does not occur in animals, probably because of the difficulty of conveying blood supplies and nervous impulses past a rotating joint while making use only of the usual biological mechanisms. In the robot there is no such restriction, and therefore there is some advantage in the use of wheels as the motive source.

If wheels are to be used, it has to be decided how many wheels are appropriate, how they shall be driven, how the speed shall be controlled, whether they shall be operated by a reversible drive, what form of braking should be used and what form of springing and tyres are appropriate. All of these points must be determined with reference to the particular range of applications being envisaged, and from an economic point of view this range of potential applications should be as wide as possible.

A mobile robot must be as manoeuvrable as possible in a small space, and it should preferably be capable of turning around without requiring a large turning circle. From this point of view it would be desirable to have three wheels, all independently steerable. However, such complication of control is unlikely to be obtainable in a small mobile robot.

While there are advantages in the use of caterpiller tracks on a robot, they would be undesirable for use on the domestic robot unless special steps were taken to prevent marking of the floor during turning. On the other hand, some form of track makes it possible for the robot to pass over obstacles which would halt a wheeled robot. As an example, a tracked robot is capable of climbing stairs, and such a device has been suggested for use on indoor invalid carriages. Tracked robots are very suitable for use over rough terrain, of course; thus for use in planetary exploration such devices might well be used. However, weight is a severe problem here, and an alternative which has been proposed is the use of very lightweight, and very flexible, wheels having a large diameter and very springy spokes One form of tyre which has been proposed for this use[48] is constructed from a mesh of 0.85 mm diameter steel piano wire crimped at intervals of 5 mm. A total of 800 strands, each 80 cm long is woven and shaped by hand and mounted on spun aluminium discs. The outer tread is made from titanium strips formed in a herringbone pattern to give abrasion resistance, while inside the mesh tyre titanium bumper rings are mounted in order to limit the amount of deflection as the tyre hits bumps. The possibilities of a robot learning cyclist have been considered by Penev[116].

Many robots will have to operate in the same environment as human beings. This environment is adapted for locomotion on two feet by walking, with one foot leaving the ground at a time. Unfortunately, this mode of movement is basically unstable; and if a robot was required to walk like a human, it would be necessary for it to be capable of learning to walk as does the human child.

Consequently, walking robots in the past have usually been merely a form of wheeled vehicle, with wheels on each foot. Neither foot leaves the ground. There seems to be little advantage in such an approach, compared with the simple use of wheels without a leg-like movement. Thring has, however, introduced a form of single walking foot for a robot which gives stable support by the intermeshing of two sets each having two rods pointing inwards[118]. This is similar in principle to the supports used for the walking dragline crane. Either set of two rod-feet alone can support the robot, while the other set is moved forward.

Another way of overcoming the problem of balance is by use of more than two legs. For example, three legs can ensure a static balance, though they are not of much help with balance when mobile unless very specially shaped feet are used, designed so that under all circumstances the vertical through the over-all centre of gravity of the robot falls within the supports.

Four legs make possible a progression using one leg at the time, the other three then ensuring balance. In general, the more legs that are used, the easier the problem of balance maintenance when moving. This method has been used in the walking dragline crane. However, at the same time the advantage of legs over wheels in obstacle-avoiding and in step-climbing is reduced as the number of legs is increased.

If a number of legs are used, so that some of them are redundant, then it is advantageous to make provision for retraction of any particular leg in the event of its failure, so that it does not limit the possibility of movement.

In the achievement of mobility it appears to be unlikely that the robot will use the human method. In particular, human feet are ill-shaped for the maintenance of balance on the ground.

Thus, all forms of locomotion on legs, wheels or caterpillar tracks, or even using the hover principle, have disadvantages, and the method to be preferred depends on the main application of the robot.

Lunokhod

It is of interest to compare some of the approximately known characteristics of the very successful Soviet mobile moon robot

Lunokhod with the approximately known characteristics of the manned U.S. vehicle Lunar Rover[60, 61].

	U.S. Lunar Rover	Lunokhod 1
Weight	210 kg	750 kg
Load	450 kg	
Length	310 cm	220 cm
Width	205 cm	214 cm
Height	115 cm	138 cm
Capacity	2.3 m^3	6.5 m^3
Wheels	4	8
Diameter	80 cm	50 cm
Drive motors one	185 W per wheel	ind. wheel motors
Steering motors	75 W fwd./rev.	skid type
Tread	180 cm	160 cm
Wheelbase	230 cm	170 cm
Journeys	3	Intermittent
Distance	36 km	8 km
Duration	12.5 h	9 lunar days
Power	Two 36 V AgZn 121 AH batteries	solar array plus batteries; radioisotope heat store

The eight wheels of Lunokhod are mounted in pairs with independent springing and individual two-speed drive. The load of the Rover consists of 180 kg per man, 60 kg of scientific apparatus and 30 kg of samples.

REFERENCES

1. Marriott, J., 'Robotugs Are on the March', *Electron. Wkly*, October 6, 7 (1965).
2. Rosen, C. A., 'Machines that Act Intelligently', *Sci. J.*, **4**, October, 109 (1968).
3. Maguire, H. T. and Arnold, W., 'Intelligent Robots: Slow Learners', *Electronics*, **40**, May 1, 117 (1967).
4. Anon., 'Gripping Thoughts of Robot Freddy', *Eng. Now*, November 27, 8 (1970).
5. Clark, J. W., 'The Mobot Mark 3 Remote Handling System', *Proc. 9th Conf. Hot Labs Equipt, November, 1961* (A.N.S.), 111.
6. Ballinger, H. A., 'Machines with Arms', *Sci. J.*, **4**, October, 59 (1968).
7. Barabaschi, S., *et al.*, 'An Electronically Controlled Servomanipulator', *Proc. 9th Conf. Hot Labs Equipt, November, 1961* (A.N.S.), 143.
8. Mancini, C. and Roncaglia, F., 'The Electronic Servomanipulator Mascot 1 of C.N.E.N.', *Alta Freq.*, **32**, 379 (1963).
9. Goertz, R. C., 'Fundamentals of General Purpose Remote Manipulators', *Nucleonics*, **10**, November, 36 (1952).

10. Ferguson, K. R., et al., 'Remote Handling of Radioactive Materials', in: McLain, S. and Martens, S. H. (eds.), *Reactor Handbook*, 2nd edn, Vol. 4, Ch. 14, 463, Interscience (1964).
11. Goertz, R., et al., 'Preliminary Report on the ANL Mark E4A Master-Slave Manipulator', *Proc. 14th Conf. Remote Systems Technol.*, *1966* (ANS).
12. Galbiati, L., et al., 'A Compact and Flexible Servo-System for Master-Slave Electric Manipulators', *Proc. 12th Conf. Remote Systems Technol., November, 1964* (ANS), 73.
13. Potts, C. W., et al., 'Transistorised Servosystem for Master-Slave Electric Manipulators', *Proc. 9th Conf. Hot Labs Equipt, November, 1961* (ANS).
14. Ballinger, H. A., 'Telechiric Devices and Systems', in: *Ministry of Technology Conference on Sea and Seabed Technology*, H.M.S.O. (1967).
15. Laymen, D. C. and Thornton, G., *Remote Handling of Mobile Nuclear Systems*, U.S. A.E.C. Div. of Tech. Inf., Book ref. TID 21219 (1966).
16. Homer, G. M., 'Mobile Manipulator Systems', *Proc. 14th Conf. Remote Systems Technol., 1966* (ANS).
17. Anon., 'A Radio Controlled Tractor', *Electron. Eng.*, **38**, July, 457 (1966).
18. Anon., 'An Automatically Controlled Farm Tractor', *Electron. Eng.*, **31**, January, 43 (1959).
19. Sutro, L. L. and McCulloch, W. S., 'Steps toward the Automatic Recognition of Unknown Objects', *Proc. IEE/NPL Conf. Pattern Recog., 1968*.
20. Feldman, J. G., et al., 'The Stanford Hand-Eye Project', *Proc. Int. Joint Conf. Artificial Intelligence, New York, 1969* (ACM), 509.
21. Helps, F. G., 'Driverless Tractors for Materials Handling', *J.Br.IRE*, **25**, March, 273 (1963).
22. Sutro, L. L. and Kilmer, W. L., 'Assembly of Computers to Command and Control a Robot', *A.F.I.P.S. Conf. Proc.*, **34**, Spring, 113 (1969).
23. Keckler, W. G. and Larson, R. E., 'Control of a Robot in a Partially Unknown Environment', *Automatica*, **6**, 469 (1970).
24. Gordon, D. A., 'Experimental Isolation of Driver's Visual Input', *Publ. Rds, Wash.*, **33**, 266 (1966).
25. Gordon, D. A., 'Perceptual Basis of Vehicular Guidance', *Publ. Rds, Wash.*, **34**, 53 (1966).
26. Biggs, L., 'Directional Guidance of Motor Vehicles', *Ergonomics*, **9**, May, 193 (1966).
27. Weit, D. H. and McRuer, D. T., 'Dynamics of Driver Vehicle Steering Control', *Automatica*, **6**, 87 (1970).
28. Penoyre, S., 'A Robot in the Driver's Seat', *New Scient.*, **50**, May 13, 371 (1971).
29. Hender, B. S., 'Recent Developments in Battery Electric Vehicles', *Proc. IEE*, **112**, December, 2297 (1965).
30. Byrne, J. V. and Lacy, J. G., 'Compatible Controller-Motor System for Battery-Electric Vehicle', *Proc. IEE*, **117**, February, 369 (1970).
31. Eadie, R. J., 'Electronic Speed Control for DC Motors', *Electron. Eng.*, **40**, January, 10 (1968).
32. Barrow, H. G. and Salter, S. H., 'Design of Low-Cost Equipment for Cognitive Robot Research', *Mach. Intell.*, **5**, 555 (1969).
33. Ernst, H. A., 'MH 1, A Computer Operated Mechanical Hand', *A.F.I.P.S. Conf. Proc.*, **21**, 39 (1962).
34. Cutler, C. C., 'Man, A Subsystem?' *IEEE Spectrum*, **4**, August, 69 (1967).
35. Nilsson, N. J., 'A Mobile Automaton', *Proc. Int. Joint Conf. Artificial Intelligence, New York, 1969* (A.C.M.), 509.

36. Nilsson, N. J. and Raphael, B., 'Preliminary Design of an Intelligent Robot', in: Tou, J. T. (ed.), *Computer and Information Sciences*, Vol. 2, 235, Academic, New York (1967).
37. Khol, R., 'The Electric Brain', *Mach. Des.*, **71**, May 29, 102 (1969).
38. Kao, H. S. R., 'A Feedback Analysis of Eye Head Angular Displacements in Human Vehicular Guidance', *Proc. Int. Symp. Man–Machine Systems, September, 1969*, Vol. 2.
39. Pout, H. W., 'The Evolution of Guided Weapons', *Aeronaut. J.*, **73**, 547 (1969).
40. Augustine, N. R. and Yates, R. M., 'The Evolution of the US Tactical Missile Program', *Aeronaut. J.*, **74**, 957 (1970).
41. Harris, G. O. and Perkins, D. J., 'A Light-Operated Gradient Controller for Agricultural Drain Laying Machines', *Electron. Eng.*, **37**, June, 364 (1965).
42. Cox, S. W. R., 'Automatic Control of Static and Field Equipment', *IEE Colloquium on Automation in Farming, January, 1969*, 1/1.
43. Brooke, D. W. I., 'Off-the-Wire Guidance for Leader Cable Vehicles', *IEE Colloquium on Automation in Farming, January, 1969*, 8/1.
44. Owen, V. M., 'Automation in the Field', *IEE Colloquium on Automation in Farming, January, 1969*, 2/1.
45. Warner, M. G. R., 'Automatic Control for the Driverless Operation of Farm Tractors', *IEE Colloquium on Automation in Farming, January, 1969*, 4/1.
46. Owen, V. M., 'Livestock Environment Control', *IEE Colloquium on Automation in Farming, January, 1969*, 6/1.
47. Warner, M. G. R., 'The Automation of Agricultural Field Work', *Proc. Istn Mech. Engrs*, **179**, pt 3H, 295 (1965) (Instn Mech. Engrs/U.K. Automation Council).
48. Anon., 'Wire Mesh Wheel Mimics Pneumatic Type', *New Scient.*, **50**, June 3, 574 (1971).
49. Rowley, W. P. and Hannaford, D. E., 'Controlled Range "Wireless" Operation of Mechanical Handling Machines', *Proc. Joint Conf. Electron, Control Mech. Handl., Nottingham, July 1971* (IERE), 145.
50. Masakasu Ejiri, et al., 'An Intelligent Robot with Cognition and Decision-Making Ability', *Proc. 2nd Int. Joint Conf. Artificial Intelligence, London, September, 1971* (Br. Computer Soc.), 350.
51. Pingle, K. K., Singer, J. A. and Wichman, W. M., 'Computer Control of a Mechanical Arm Using Visual Input', *Proc. I.F.I.P.S.*, 1563, North-Holland (1968).
52. McCarthy, J., et al., 'A Computer with Hands, Eyes and Ears', *Proc. Fall Joint Computer Conf.*, **33**, 329 (1968).
53. Tenenbaum, J. M., et al., 'A Laboratory for Hand-Eye Research', *Proc. I.F.I.P.S.*, Ljubljana (1971).
54. Pingle, K. K., 'Visual Perception by a Computer', in: *Automatic Interpretation and Classification of Images*, 277, Academic Press (1969).
55. Hueckel, M., 'An Operator which Locates Edges in Digitized Pictures', *J. Ass. Computer Machinery*, **18**, January, 113 (1971).
56. Pingle, K. K. and Tenenbaum, J. M., 'An Accomodating Edge Follower', *Proc. 2nd Int. Joint Conf. Artificial Intelligence, London, September, 1971* (Br. Computer Soc.), 1.
57. Falk, G., 'Scene Analysis Based on Imperfect Edge Data', *Proc. 2nd Int. Joint Conf. Artificial Intelligence, London, September, 1971* (Br. Computer Soc.), 8.

58. Feldman, J. A. and Sproull, R. F., 'System Support for the Stanford Hand–Eye System', *Proc. 2nd Int. Joint Conf. Artificial Intelligence, London, September, 1971* (Br. Computer Soc.), 183.
59. Feldman, J. A., et al., 'The Use of Vision and Manipulation to Solve the "Instant Insanity" Puzzle', *Proc. 2nd Int. Joint Conf. Artificial Intelligence, London, September, 1971* (Br. Computer Soc.), 359.
60. Zimmerman, M. D., 'Apollo 15: Design for Moon Mobility', *Mach. Des.*, **43**, July 22, 20 (1971).
61. Anon., 'Lunar Rover Expands the Frontiers of the Moon', *Mach. Des.*, **43**, July 22, 57 (1971).
62. Anon., 'Air Cushion Handling at Cranfield with Disc Skirt', *Mater. Handl. News*, June, 60 (1971).
63. Warner, M. G. R., 'Driverless Farm Tractors', *Electron. & Power*, **17**, August, 308 (1971).
64. Kay, A. C., 'Manipulators as Terminal Devices', *Proc. IEEE Int. Comput. Group Conf., Washington, June, 1970*, 290.
65. Barrow, H. G., 'The Development of a Real World Interface', *Proc. Conf. Man–Computer Interaction, NPL, September, 1970* (IEE), 89.
66. Anon., 'Electronic Orderlies Hasten Hospital Automation', *Electronics*, **44**, September 13, 36 (1971).
67. Anon., 'Electronics in the Driver's Seat', *Electronics*, **44**, November 22, 89 (1971).
68. Interian, A., Allen, W. and Kugath, D. A., 'A Remote Manipulator System for Space Applications', *Proc. Int. Symp. Man–Machine Systems, Cambridge, September, 1969* (IEEE).
69. Blackmer, R. H., Interian, A. and Clodfelter, R. G., 'The Role of Space Manipulator Systems for Extra-Vehicular Tasks', *Proc. 2nd Natl Conf. Space Maintenance and Extravehicular Activities, Las Vegas, August, 1968*.
70. Bradley, W. E., 'Telefactor Control of Space Operations', *Astronaut. Aeronaut.*, **5**, May, 32 (1967).
71. Sheridan, T. B. and Ferrell, W. R., 'Human Control of Remote Computer-Manipulators', *Proc. Int. Joint. Conf. Artificial Intelligence, New York, 1969* (A.C.M.), 483.
72. Ferrell, W. R., 'Remote Manipulation with Transmission Delay', *Trans. IEEE*, **HFE6**, September, 24 (1965).
73. Ferrell, W. R., 'Delayed Force Feedback', *Hum. Factors*, **8**, October, 449 (1966).
74. Ferrell, W. R. and Sheridan, T. B., 'Supervisory Control of Remote Manipulation', *IEEE Spectrum*, **4**, October, 81 (1967).
75. Cole, I. S., 'An Experiment in Robot Tool Using', in: *Systems for the Seventies*, 224, IEEE (1970).
76. Fenton, R. E., et al., 'One Approach to Highway Automation', *Proc. IEEE*, **56**, April, 556 (1968).
77. Bender, J. G. and Fenton, R. E., 'A Study of Automatic Car-Following' *Trans. IEEE*, **T18**, November, 134 (1969).
78. Gazis, D. C., et al., 'Non-Linear Follow-the-Leader Models of Traffic Flow', *Op. Res.*, 545 (1961).
79. Levine, W. S. and Athans, M., 'On the Optimal Error Regulation of a String of Moving Vehicles', *Trans. IEEE*, **AC11**, July, 355 (1966).
80. Peppard, L. E. and Gourishankar, V., 'Optimal Control of a String of Moving Vehicles', *Trans. IEEE*, **AC15**, June, 386 (1970).
81. Peppard, L. E. and Gourishankar, V., 'An Optimal Car-Following System', *Trans. IEEE*, **VT21**, May, 67 (1972).

82. Szekely-Doby, S., 'Research on Vehicles without Conductors for the Improvement of Highway Schemes', *Automatism*, **17**, April/May, 140 (1972).
83. Pawula, J., 'Moving Material with Driverless Trains', *Automation*, **19**, June, 69 (1972).
84. Anon., 'Radar Controlled Car: A Step toward Automated Highways', *Ind. Wk*, **172**, February 14, 65 (1972).
85. Anon., 'Optics for Readouts (Ferranti Light-Guided Car)', *Electron*, April, 11 (1972).
86. Anon., 'London Traffic under the Magic Eye', *Control & Instrum.*, **4**, July/August, 9 (1972).
87. Blakelock, J. H., *Automatic Control of Aircraft and Missiles*, Ch. 6, Wiley (1965).
88. Dymond, E. G., 'Measurements in the Upper Air by Radio Sonde', *Research*, **3**, 345 (1950).
89. Dymond, E. G., 'The Kew Radio Sonde', *Proc. Phys. Soc.*, **59**, 645 (1947).
90. Jones, F. E., et al., 'The Radar-Sonde System for the Measurement of Upper Wind and Air Data', *Proc. IEE*, **98**, pt 2, August, 461 (1951).
91. Henslow, M., *The Miracle of Radio*, 104, Evans (1946).
92. Baynton, E. W., et al., 'Jindivik-Radio Controlled Aircraft', *Proc. IRE Aust.*, **17**, August, 267 (1956).
93. Anon., 'A Drone Air Force for Combat Being Designed by Two Teams', *Electron. Des.*, **20**, February 3, 26 (1972).
94. Aronson, R. B., 'RPVs: The End of Manned Military Flight?', *Mach. Des.*, **44**, April 20, 20 (1972).
95. Anon., 'Automatic Craft Brings Back Oldest Rock Samples from Moon', *New Scient.*, **53**, March 9, 545 (1972).
96. Barkan, R., 'The Robot Airforce Is About to Take Off', *New Scient.*, **55**, August 10, 280 (1972).
97. Blackmer, R. H. and Clodfelter, R. G., 'The Application of Remote Manipulators in Space', *Nucl. News*, **12**, March, 40 (1969).
98. Interian, A. and Kugath, D., 'Remote Manipulators in Space', *Astronaut. Aeronaut.*, **7**, May, 40 (1969).
99. Hunt, C. L. and Linn, F. C., 'The Beetle', *S.A.E. Jl*, **70**, September, 52 (1962).
100. Jones, D. G., 'MRMU in Case of Radioactive Trouble', *Mech. Eng.*, **86**, May, 29 (1964).
101. Spielrein, R. E., 'Some Modern Prosthetic and Orthotic Trends and Developments Seen as a Challenge to the Engineering Profession', *J. Inst. Engrs Aust.*, **41**, June, 73 (1969).
102. Newman, N. and Tait, K. E., 'Manipulators: A Survey', *Elect. Eng. Trans. Inst. Engrs Aust.*, **EE8**, April, 1 (1972).
103. Johnsen, E. G., 'Man, Teleoperators and Robots; An Optimum Team for Space Exploration', *J. Spacecraft Rockets*, **9**, July, 554 (1972).
104. Heyes, M. P. and Ashworth, R., 'Further Research on Car-Following Models', *Transportn Res.*, **6**, September, 287 (1972).
105. Masakazu, E., et al., 'Prototype Intelligent Robot that Assembles Objects from Plan Drawings,' *Trans. IEEE*, **C21**, February, 161 (1972).
106. Anon., 'Industrial Robot', *Mach. Des.*, **44**, January 13, 39 (1972).
107. Barrow, H. G., et al., 'Tokyo-Edinburgh Dialogue on Robots in Artificial Intelligence Research', *Computer J.*, **14**, February, 91 (1971).
108. Fenton, R. E., 'Automatic Vehicle Guidance and Control—A State of the Art Survey', *Trans. IEEE*, **VT19**, February, 153 (1970).
109. Hajdu, L. P., et al., 'Design and Control Considerations for Automated Ground Transportation Systems', *Proc. IEEE*, **56**, April, 943 (1968).

110. Bender, J. G., et al., 'An Experimental Study of Vehicle Automatic Longitudinal Control', *Trans. IEEE*, **VT20**, November, 114 (1971).
111. Anon., 'Adaptive Computer may Control Cars', *Electronics*, **45**, September 25, 6E (1972).
112. Rosen, C. A. and Nilsson, N. J., 'An Intelligent Automaton', *IEEE Int. Conv. Rec.*, pt 9, 50 (1967).
113. Fu, K. S., 'Learning Control Systems, Review and Outlook', *Trans. IEEE*, **AC15**, April, 210 (1970).
114. Keckler, W. G. and Larson, R. E., 'Control of a Robot in a Partially Unknown Environment', *Automatica*, **6**, May, 469 (1970).
115. Fu, K. S., 'Learning Control Systems and Intelligent Control Systems; An Intersection of Artificial Intelligence and Automatic Control', *Trans. IEEE*, **AC16**, February, 70 (1971).
116. Penev, G. D., 'Certain Problems in Adaptive Control', *Soviet Phys. Dokl.*, **16**, No. 6, 422 (1971).
117. Tomovic, R., 'Robots for the Exploration of the Hostile Environment', *Proc. 4th I.F.A.C. Symp. Automatic Control in Space, Dubrovnik, 1971*.
118. Thring, M. W., 'The Robot Age', *Engineering, Lond.*, **209**, February 6, 128 (1970).
119. Mayhan, R. J., et al., 'Reference Signal Generation for Synchronous Longitudinal Control', *Proc. S.E. Symp. Systems Theory, University of Kentucky, 1972*.
120. Maughan, R. J., et al., Longitudinal Reference Signal Generation for Automatic Vehicle Control', *Proc. IEEE*, **60**, November, 1454 (1972).
121. Brooke, D. W. I., 'The Automation of Field Cultivation', *I.E.E. Colloquium On Control and Automation in Agriculture, 1973, March 20*.
122. Hawley, A. E., et al., 'Electronic Packaging Techniques for Surveyor Lunar Spacecraft', *IEEE Int. Conv. Rec.*, **11**, pt 6, March, 157 (1963).
123. Vinogradov, A., 'Luna 20 Samples from the Moon', *Geotimes*, **17**, October, 16 (1972).
124. Michie, D., et al., Vision and Assembly as a Programming Problem', *Proc 1st Conf. Industrial Robot Technology, Nottingham, March, 1973*, 185.
125. Falk, G., Interpretation of Imperfect Line Data as a Three-Dimensional Scene', *Artificial Intell.*, **3**, Summer, 101 (1972).
126. Fikes, R. E., et al., 'Learning and Executing Generalised Robot Plans', *Artificial Intell.*, **3**, Winter, 251 (1972).
127. Salter, S. A., 'Arms and the Robot', *Edinburgh Univ. Bionics Reports*, No. 9, April (1973).

8

Robot limbs

Practical robot arms and hands

All over the world at the present time, work on robot hands and arms is being carried on with various applications in mind. This work can be divided up roughly into a number of different categories:

1. Remotely controlled arms and hands for use in environments potentially hazardous to a human being.
2. Remotely controlled arms and hands designed to increase the effective strength of the human operator, or to increase or modify the effective size of the human operator.
3. True robot arms and hands having inbuilt inherent automatic self-control of movement, guided only by certain basic design principles.
4. Prosthetic devices for use by limbless people.

Developments in all of these fields are of interest and of use to the research worker in the field of robotics.

It has been estimated that in the nuclear industries alone many hundreds of static master–slave remote manipulators are in use by industrial workers as well as by scientists[1, 2]. However, many of these employ only straightforward mechanical control linkages manipulated by the human operator, who can see the results of his actions either directly or on television and so can modify his actions if required as a result of the visual feedback[2].

Remotely controlled manipulators[3, 4, 59-69, 91]

Because various environments are potentially hazardous to human beings, it has been necessary to develop machines suitable for remote manipulation, controlled by a human operator who can remain in a safe environment. Developments of this nature are of great interest to the robotics investigator, because they demonstrate the real need for robots, and because the forms of mechanism which have been developed might well give a guide to the production of automatically controlled limbs for truly independent robots.

The problems encountered with such remote manipulator systems are largely caused by the effects of variations in the moments of inertia of the loads. The introduction of reduction gearing between drive motor and load can help to mask such effects, but then the inertia of an electrical drive motor can itself become of greater importance. The human operator then experiences an inevitable feeling of annoyance at the delayed response of a slave unit. For such reasons electro-hydraulic operation is often preferred to the straightforward use of electric actuating motors.

In some human-controlled devices it has been found useful to provide the human operator with a tactile form of feedback in addition to the usual purely visual feedback. The operator then has a sense of touch. This feature is sometimes known as force reflection. An additional advantage of such a system is that the operator is less likely to apply an excessive force to any of the movements and so to cause damage. With any such system safety precautions are very necessary to avoid operator injury in the event of a fault. A high speed of response of the movement is necessary if physical fatigue of the human operator is to be avoided.

Factors such as the reach of a manipulator controlled by a human, the load and the capacity to hold the load continuously can easily be made to exceed the capability of the unaided human. However, while simple tasks can be carried out nearly at the same speed as the unaided hand, more complex tasks can only be carried out slowly, if at all. In general, the speed of a human-manipulator combination is only about one-eighth of the speed of the unaided human performing the same task. This limitation carries the implication that we must expect our early independent robots to be as slow as this, or perhaps even slower. However, because of the lack of fatigue and need for sleep in the independent robot, it will often be possible to accept a slow speed of operation, provided little or no human supervision is necessary, since a given task can be carried on for 24 hours a day.

Early human-controlled manipulators were only required to be capable of moving a radioactive chemical sample from one point to another. Simple hydraulically operated tongs were sometimes used. However, this form of manipulator was not very rigid and it gave a jerky response. Only small loads could be handled.

In modern form simple manipulators can be very versatile.

Consider for example the MiniManip produced by Programmed and Remote Systems Corporation. This complete unit weighs only about 7 kg. It incorporates a horizontal arm which has ±45° of freedom, with a vertical operating arm at one end and a vertical operator's arm at the other end, the horizontal arm being about 90 cm long and the vertical arms being each about 60 cm long. The entire arms can be manipulated to turn 360° in a vertical plane and ±60° from the vertical. The forearm can be twisted ±170°, while the tongs at the end can be swung 45° upwards and 135° downwards and twisted ±210°. The tong fingers at the end of the arm can be opened to about 7.5 cm, and there is a ratchet type of lock to maintain the grip. A load of about 2.25 kg can be manipulated and a volume of 0.85 m^3 can be covered.

Another version of this manipulator has a true Z motion, in which the operating arm moves upwards if the operator's arm is moved upwards, so repeating the natural motion of the operator, though, of course, a weight increase is then necessary, to about 12.5 kg. These are only the simplest manipulators of a range, and it is quite possible, for example, to amplify the vertical movement of the operator's arm by 2:1 and to provide an increased load capability or grip capability.

A study of straightforward manipulators such as these can give a useful guide to the requirements not only for powered manipulators but also for true independent robot movements.

The limitations of the purely mechanical form of manipulator are:

1. The master and the slave must normally be coupled mechanically to each other and to the mechanical base upon which the operations are to be carried out.
2. The volume of space which can be covered by the manipulator is very limited, and there is, in general, no absolute freedom of motion.
3. Since it is not usually practicable to arrange for multiplication of the force imposed by the operator, it is sometimes necessary for the operator to use two hands to carry out an operation which is nominally single-handed, and there is a possibility of operator fatigue.
4. The speed of operation which can be achieved is much slower than is that of the unaided human, being typically only one-eighth of that obtained with the hand alone.

The time taken for a human to lift off various weights of objects after the initial grip varies. Published information[27] on this would seem to be fitted by the equation

$$\text{time to lift-off} = 0.3 + \frac{\text{weight of object (kg)}}{1.8} \text{ seconds}$$

Faster response is obtainable with artificial devices.

Powered manipulators[5, 22, 23]

It is possible to add electrical or hydraulic operation to the simple forms of manipulators described above. Such forms are very suitable for the transport or the reorientation of objects. Indeed, the normal manually controlled electric crane is an elementary form of such a manipulator, particularly when it is fitted with an automatic grab mechanism or one which can be controlled by the human operator. All such controls have limited speed.

However, such a 'unilateral' form of control is found to be very limited in its possible uses. The only form of feedback to the operator in such cases is visual, and occasionally also aural. If a fine control of motion is to be achieved, it is found necessary also to add some form of mechanical or 'touch' feedback to the hand, and perhaps also to the limbs, of the human operator.

Many different types of such 'bilateral' controlled manipulators have been produced since about 1949, the incentive being mainly the necessity to work in hostile environments such as areas of high nuclear radiation or, more recently, in an underwater environment. Such devices clearly have additional application to remotely controlled work to be carried out in space but controlled by a human operator on earth, though here the additional complications introduced by the inevitable transmission delays are also encountered[28].

The simplest form of feedback to the operator, used on some manipulators, has been a graduated 'touch' dial, sometimes supplemented by an audio indication so that the operator need not watch the dial.

The early bilateral manipulators were only capable of handling a small load of about $\frac{1}{2}$ kg. Nevertheless, they demonstrated the utility of this approach, which was later developed through various models until a load capacity of a few hundred kilogram has been achieved.

The speed of operation achieved with the motor controls in manipulators, corresponding to about 10 Hz maximum, appears to be sufficient for many purposes. Indeed, it corresponds to the capability of the unaided human. However, limitation of the feed back frequency response to this frequency range proves to be a disadvantage, and it limits the speed of operation which can be achieved by a human controlling a manipulator.

It has been pointed out by Goertz that, when using a normal hand tool, the human receives a great deal of sensory information from the tool through his hand, and it is therefore desirable that the feedback frequency response should be increased in both the remote human-controlled manipulator, and in the self-controlled robot[6].

There is a need for work on such high-fidelity feedback systems, having a frequency response up to 100 or even 1000 Hz. However, the introduction of such sensory feedback is likely also to introduce problems due to the feedback of false information introduced by the deficiencies of the system[30].

An American arrangement used in experiments on nuclear-powered flight was the Minataur, which had a ball-shaped body with five articulated arms, three carrying T.V. cameras. It hung from a mobile roof-crane.

It is of interest to give a typical specification for an electrically powered manipulator which has the same over-all shape as the human arm.

HAND (interchangeable)
   Open–close speed   1 cm/s
   Force   0–350 N
   Maximum opening   6.5 cm

WRIST Pivot torque   17 Nm
   Pivot speed   2 rev/min
   Pivot arc   320°
   Rotational torque   3.5 Nm (continuous rotation is possible)

ELBOW Pivot torque   55 Nm
   Pivot speed   2 rev/min
   Pivot arc   320°
   Distance from wrist to elbow   40 cm

SHOULDER Pivot torque   100 Nm
   Pivot speed   2 rev/min
   Pivot arc   228°
   Distance from elbow to shoulder   38 cm

BODY Rotation torque   20 Nm
   Rotation speed   3 rev/min (continuous rotation)
   Lifting capacity   25 kg

144 Robot Limbs

Electrical master–slave manipulators[7-9, 53, 56]

The use of electrical drive with manipulators brings several advantages. For example, an extreme mobility of the slave can be achieved, while the master control can be so positioned that the best possible visibility is obtained. It is very easy with electrical drive to have the master control force either much less than or much greater than the slave force, and the distance–movement ratio can similarly be varied, while these two ratios can be made quite independent of each other. Consequently, the master and slave arms can have very different sizes and shapes. This has great advantages where the load to be manipulated is either very large or very small. The former is the obvious case, and it seems to have been more exploited, but the possibility of construction of microscopic master–slave manipulators holds great promise for future work—for example, on the development of integrated circuits.

It is important that some robot units, and some slave units, should be easily repaired by similar units. For example, this is essential for space applications, and it is very useful in nuclear applications where the slave unit can be contaminated. Electrically controlled units have been designed which can be disassembled and repaired by similar units. Such work is likely to be even more extended in the future as the number of robots in use increases, since it is very desirable to avoid the necessity for human labour for any routine maintenance operations.

It is found in practice that the human operator achieves a reasonable dexterity with these devices after only 1 hour of use, though it can take several weeks or months of use to master the required movements, since only two fingers are available on the slave hand. It seems that the first obstacle in human thought in acquiring dexterity is that the human tends to think always in terms of his own required actions at the controls, and it is necessary for him instead to think in terms of the required action of the tongs of the manipulator. This problem will not be encountered with the untrained robot, since it is mainly a problem of preconceived ideas.

Experiments have shown that with such manipulators quite high load inertias and frictions can be tolerated, as can a relatively low stiffness of coupling between master and slave. The installation of clipping devices which limit the maximum signal and therefore the maximum force has been found to lead to a significant increase of reliability of such systems.

One problem encountered with manipulators of the closed-loop type is the difficulty of achievement of stability of the operating servo systems despite the varying nature of the load. This leads to a servo control system which is basically nonlinear. In work on such systems having effectively varying parameters such as time constants the present writer has been able to demonstrate the applicability of reasonably simple design methods in some cases—for example, where the main load time constant varies over an effective 12:1 range[10, 11], as mentioned in Chapter 6.

Control-box-operated manipulators

There are various current applications in which remotely operated manipulators are controlled by knobs on a control box rather than by a more direct copying of the motion of a human.

A typical unit, the PAR3000, has parallel-jaw vice-grip hands with a finger travel of about 12.5 cm capable of applying a maximum grip force of about 900 N with a closure moving velocity of about 7.5 mm/s. Other forms of hand can be interchanged easily. The gripping force is controlled by an electric clutch, the maximum force being set by a knob on the control console. The front section of the wrist assembly is easily removable, together with the drive motors which it contains, and the internal parts are then accessible.

The lower and upper arm sections are constructed from a continuous box-like structure, since this combines the advantages of rigidity, sealing and ease of disassembly. The sealing is by means of 'O' rings. Externally adjustable chain tighteners are mounted inside the arm, and access plates are provided for maintenance. The roller chains transmit power from the motors driving the wrist, elbow and shoulder joints. The motors, together with their protective slip clutches, are mounted in the shoulder housing, and the clutch setting can be adjusted by removal of a cover on the shoulder. Continuous shoulder rotation is provided by means of a shoulder motor, again fitted with a slipping clutch.

A control console measuring about 38 cm × 20 cm × 15 cm is used by the operator, centre-zero sliding knobs being used for control of linear motions and rotary knobs for rotary motions. The control circuitry and the power supplies for the electric motors are mounted in a cabinet mounted on castors. Magnetic amplifier and thyristor types of stepless controls are used, together with overload protection. The total power requirement is about 1 kVA. A socket is provided in the wrist assembly, so that power is available for external electric tools to be used by the manipulator.

A sideways motion of the wrist pivot can be provided if required, and the entire manipulator can be mounted on a traversing bridge system, with a telescoping arm assembly. A hook capable of carrying

146 *Robot Limbs*

a heavy load is provided on the shoulder housing, and this can easily be reached by the hand of the manipulator.

The manipulator is completely sealed, so that it can be washed down, and all of the electric wiring is internal. Similar devices are manufactured for underwater use with a claimed unlimited depth capability. The inside of such a device is filled completely with a lubricating fluid, with a depth compensator to keep the pressure slightly above the ambient pressure. Hydraulic motors are used for the drives, with a nominal hydraulic supply pressure of about 12 000 kN/m².

Thus the technology of the production of advanced manipulator arms and hands is progressing, and there is no doubt that such devices will be directly applicable to the fully mobile robot. The immediate requirements for such applications are a reduction of weight and of power supply requirements. This is 'natural selection'.

Prosthetic arms and hands[57, 58, 72-88]

There has been a great deal of work all over the world on prosthetic devices, and this has been intensified by the thalidomide tragedy. A typical recent example is the artificial arm and hand developed by the N.E.Co. of Montreal[25]. Although this arm and hand assembly weighs little more than 1 kg, it is capable of being used to lift a mass of 10 kg using the hand and the flexing elbow. There are four basic movements with this prosthesis, including forward and backward shoulder movement, a natural elbow movement, a wrist movement through an angle of 180°, and finger and thumb movements of the hand.

The main power supply is a 12 V nickel–cadmium battery which carries sufficient charge for 1 day's operation before recharging. A high-speed electric motor drives a hydraulic pump which provides an odourless hydraulic fluid at high pressure via plastics tubing to the hydraulic actuators mounted in the arm. Main control can come directly from the nerves via myo-electric signals.

A worm gear and lever system operates the hand via miniature instrument-type roller chains which simulate human tendons. Grip can be between the fingers and the thumb; alternatively, the thumb can be released and the fingers can be used to grip alone.

Such a device not only offers a marvellous hope for prosthetic users but would also seem to have direct applications to the robot.

With a robot arm there is no necessity for it to be jointed as with a human or a prosthetic arm. It has sometimes been suggested that the jointed arm or leg has the advantage over a telescopic form that the retracted length, or distance from shoulder to hand, need be only of the order of one-sixth of the fully extended length.

However, a telescoping form of arm can have a large ratio of extended to retracted length provided that it has enough sections, and there is the additional possibility of retraction right into or even straight through the body of the robot. This approach is in fact used with the industrial robot arms which are to be described in the next chapter.

Nevertheless, there is an advantage in the jointed form. This is that a bending arm can reach around obstacles in order to enable the hand to work on inaccessible objects. The degrees of freedom are increased, though, of course, this is at the expense of the necessary increased complexity of control.

The actuation of prosthetic devices is discussed in Chapter 5.

Exoskeletons[94]

There has been some work on machines intended to be worn by a human and so to be used to amplify his strength. Such a machine is known as an exoskeleton. In the development of such devices the possible basic motions of the human body were first studied. The human can carry out some 20 or 30 different controlled actions simultaneously.

Investigations started in the mid 1950s at Cornell Aeronautical Laboratories indicated that an adaptable exoskeleton would have to have some 35 pinned joints and one sliding joint as a minimum if it was to be wearable without inconvenience by a human.

A design study was then made of the forces and sizes and types of actuator necessary for such a device. It was found that for a load of about 4500 N applied in any direction to the hand, it was necessary to have the cross-sectional area of the elements measuring 2.5 cm × 5 cm between the spine and the elbow and 2.5 cm × 2.5 cm between the elbow and the hand.

In order to avoid the necessity for geared electrical drives or for the conversion of linear motion to rotary motion with large angular displacements, special rotary hydraulic actuators were chosen. These had to be of 10 cm diameter and 10 cm length in order to support the required load, the pressure drop being about 200 kN/m^2 across the actuators. It was suggested that a greater pressure drop was desirable in order to reduce the weight and the size of the equipment. Forces were to be reflected to the operator only if they exceeded a certain frequency.

Balance sensors

The human body, when standing or walking on two feet, cannot be regarded as a stable structure. Consequently, there is a necessity for very sensitive detection of rotational movements in any direction, so that corrective action can be taken[92].

In the human being the sensing of these rotational movements is provided by the semicircular canals of the inner ear. The signals from the semicircular canals can also be used to provide signals to the eye muscles, so that they can compensate for the inevitable movements of the head and of the body. In this way a stabilised image can be maintained on the retina.

The semicircular canals comprise a smooth-bored tube, of about 0.28 mm diameter in the human being forming a closed circuit completed by a chamber called the ampulla. Inside the ampulla chamber is a watertight flap, known as the cupula. Deflections of the cupula, caused by liquid pressure in the small-bore tube, cause the transmission of neural signals from the hair cell.

The viscous resistance to the flow of liquid (or endolymph) is linearly dependent on the flow velocity, since the canal tube has a smooth bore. The springiness of the cupula flap affects the liquid movement by reacting on the viscous friction and on the fluid inertia of the liquid. There are three mutually perpendicular semicircular canals in each of the inner ears. Although the three are connected by a common compartment called the utricle, the intercoupling does not seem to be an important factor in the operation of the semicircular canal system.

The over-all response of the semicircular canal system to the angular velocity of the head movement takes the form

$$\frac{\text{output deflection}}{\text{angular velocity of skull}} = \frac{ksT}{(1+sT_1)(1+sT)}$$

Here the values of T_1 and of T are very widely spaced, so that the over-all response becomes very much like that of an electronic R–C coupled amplifier. Typical values of time constants are $T \approx 10$ s; $T_1 \approx 0.005$ s. Consequently, for fast movements the output deflection is proportional to the angular velocity of the skull. The response falls off for very rapid movement, however. The long time constant in this system presumably makes it possible for the system slowly to reset to zero at the end of a movement.

The presence of some form of angular velocity or angular acceleration sensor in the robot head would be very desirable, provided that such a sensor could act into a learning system, so that it would not be necessary to attempt to set up the control system for correct operation.

A robot which is mounted on wheels for locomotion would only need two sensors at right angles, but a robot on some form of legs (for example, a climbing robot, or one which could move vertically up stairs) would require at least three sensors performing the function of the human balance sensors, Suitable sensors have been developed as a result of work on inertial navigation systems, though smaller and less accurate sensors are required for landborne robots.

Other methods

Recent work on devices such as guided weapons has brought about an interest in equipment which is suitable, in the robot, for detecting changes of posture or of balance. For example, such sensors are required in gyro reference systems and in radio aerial and camera stabilisation equipment. A typical pendulum-type device for this purpose gives a maximum output of 25 V and can detect tilts up to an angle of 1° in each axis. The required excitation is 50 V at 400 Hz. The resistance element used in such sensors is electrolytic[12]. While this ensures a large output and a high sensitivity, it causes a large change of output with temperature. Also, since the viscosity of the electrolyte is also temperature-sensitive, the damping of the pendulous element is very sensitive to temperature, a variation of 10:1 not being unknown if the temperature variation is large.

These problems are avoided in the electromagnetic form of sensor, since the temperature sensitivity can be made small and the liquid can be chosen purely from considerations of mechanical damping. However, the output voltages obtainable are much less than those from the electrolytic type.

Other electrolytic devices have used the effect of movement of a bubble position in an electrolytic liquid, usually to provide a vertical reference for gyro systems.

The writer has been associated with electromagnetic sensors for two-axis control of machine tools. These give good repeatability but are rather bulky for robots.

The complex methods which have been used in aircraft and guided missiles for attitude determination and for inertial navigation would seem to be ideal for robot use. However, such an approach probably provides a needless accuracy and excessive cost. Additional disadvantages are the length of time required for starting up and the relatively heavy power consumption.

A pendulum having a very long period can be used, and one

150 *Robot Limbs*

possibility is a flywheel having a slightly eccentric centre of gravity. Periodic times in excess of 30 s would be required if errors due to horizontal accelerations were to be minimised, and the bearings would have to be shockproof and yet have a very low friction[19].

The values of angular accelerations involved are of the order of 1 rad/s^2. It has been suggested that the semicircular canals can be considered to act as an integrating angular accelerometer. In the human the second derivative of vertical velocity also appears to be important, as witness the strange feelings produced by traversing a hump-backed bridge in an automobile, or by travelling in a badly designed elevator.

As mentioned later, an experimental semicircular canal has been made.

The robot storekeeper

Even in a factory involved in quantity production, the operation of the stores has a large effect on the over-all economics of the operation of the plant. Indeed, in recognition of this the writer evolved in industry[31] a method of control of all of the processes in a factory which is based firmly on the method of coding used for goods in the stores.

In this system, which was used in the manufacture of industrial electronic equipment, all possible items in the stores and all paperwork in the factory, including, for example, drawings, were given a code number consisting of a two-letter identifying code followed by a number, of three figures in a medium-sized factory, which identifies the item within the identifying category. Such a system can be most successful if it is used for all production, sales and purchasing control. More elaborate schemes have been used, involving all-figure codes based on, for example, shape, though these do, in general, require some special staff for the administration of the scheme.

Now the basic principle of such a stock control system is that the stores is really the heart of the factory. Recognising this, Thring has proposed the use of a robot storekeeper[29]. This would have the advantages that it is tireless and free from boredom; can receive and remember a long string of instructions from a computer; can signal back the state of the stores to the computer; and do all this with a very low error. At the same time it can have the strength of a forklift truck and the skill of a human hand.

The tasks to be performed by such a robot storekeeper are to receive the goods incoming—for examples, from lorries; to open the parcels and to read the labels; to determine the correct stores bin—

for example, by reference to a central computer; and then to store the goods. In a similar way the robot must be able to issue items from the stores and also to carry out periodic stocktaking. Thring has suggested a possible design for a robot arm mounted on a mobile wheeled trolley and fitted with several different hands. This robot should be capable of locating the required bin, pulling it out, picking up the required objects or packets, placing them in a container and then sealing and stacking the container. Either tactile or optical sensing could be used, and while dead reckoning of the position of the robot might be satisfactory for early applications, a continuous sensing system will probably be required for later robot position determination.

Stacking cranes controlled by individual mini-computers are being used in automated warehousing[55].

A French system[90] known as Robot Systeme uses a Transferobot transporter and a Transrobot. The former can carry the latter and it controls the Transrobot via a flat, reeled cable. The Transferobot is in turn linked to a central control console by a reeled, flat cable. Magnetic detector devices are used for location.

Human walking[13, 14]

It is of interest to consider the performance of the human in walking so that a comparison with a walking robot can be made.

If as a first approximation the human leg is considered to be a stiff cylinder with a length of, say, 90 cm, a diameter of 12 cm and a density of 1.1 g/cm³, then the volume can be taken as about 10 000 cm³ and the weight as about 11 000 g. This is approximately correct for a male adult weighing about 70 000 g in all.

The equation of motion of such a simplified leg is given by

$$\frac{d_2 \theta}{dt^2} = -1.5 \times \frac{g}{\text{length}} \times \sin \theta$$

where θ is the angle of deflection of the leg from the vertical and g is the gravitational constant. For small angles of deflection, it can be assumed that $\sin \theta \approx \theta$ and then the undamped natural frequency of swing of the leg becomes approximately equal to

$$\frac{1}{2\pi} \left\{ 1.5 \times \frac{g}{\text{length}} \right\}^{\frac{1}{2}}$$

This gives about ⅔ swings/s or, in a more military notation, about 80 paces/min.

Since this is the natural frequency of swing, faster walking requires the expenditure of additional energy. It has been suggested that the wearing of a spring between the legs to increase the natural frequency can increase the rate of walking without increasing the energy expenditure excessively.

Attempts have been made to extend this form of analysis to take into account the effect of the rise and fall of the centre of gravity during walking because of the effect of the human feet.* In addition, attempts have been made to demonstrate the existence of an optimum speed of walking in the human being to give a minimum of energy expenditure.

In the human being the measurements of the legs and feet have to be accepted as fixed. Minor changes can be made (for example, by the fitting of special boots) but only if it is accepted that a weight penalty is to be incurred.

In the robot there are no limitations, and it will be possible to design the action of robot legs in order to achieve the optimum of efficiency for the task which the robot is required to perform.

The mechanical output of a human being is of the order of 375 W over a period of 30 min, and it can be up to 1125 W for very short times.

The literature on the muscle action and the energy expenditure in human walking has been reviewed by Grieve and Cavanagh[34], while the current knowledge of the relative phasing of the action of the various leg and foot muscles during human walking has been reviewed by Paul[35]. McKenzie[37] has considered the action of the knee joint in walking, while human gait characteristics, including those involved in the descent of stairs, have been studied by Contini and others[38, 93].

Frank[15-17, 24] has pointed out some very interesting facts about locomotion on legs. The method of control required is velocity-dependent in that while little or no dynamic sensing is required at a low velocity, information about the state of the body is required when the velocity is increased. The control problem must be considered both as continuous and as discrete, since the legs must leave the ground and re-establish contact. In order to control the six degrees of freedom of the body, at least six independent controls are required.

In a theoretical study of the possibility of building prosthetic legs requiring only orders such as 'Forward', 'Stop', Turn left', etc.,

---

* The relationship has been expressed as

$$\text{vertical lift} = \frac{(\text{pace length} - \text{foot length})^2}{8 \times \text{leg length}}$$

Witt[19] has suggested that the feet during walking simply provide a bearing surface and a means to push off using the ankle, and they do not perform a stabilising function as when standing still. If this is so, then the size of the feet is unimportant. Such thoughts lead to a walking arrangement with a rotary actuator at the hip, and a telescopic leg driven by a linear actuator. The ankle-joints need only be locked when standing still.

For the telescopic actuator the conventional hydraulic cylinder controlled by a valve is inefficient. Most of the telescopic action of the leg takes place when the foot is lifted from the ground and there is little opposing force. What is needed is an actuator in which the effective surface area of the piston increases if the opposing force exceeds a certain value.

As a preliminary to the completion of a two-legged machine capable of walking, Hall and Witt have successfully demonstrated a device which can stand on its two legs and can march on the spot[32]. The legs are not bent, but are extended and contracted by linear actuators, hydraulically operated.

After early attempts to use a long-period pendulum formed by a very slightly eccentrically mounted flywheel[19], gyroscopes were adopted to measure the angles of tilting, with angular pick-off devices mounted on the gimbals. Angular velocity was measured by differentiation of the signal, followed by a 10 Hz low-pass filter to remove the high-frequency noise. However, because a gyroscope requires a long run-up time and consumes power, an alternative device based on the operation of the human semicircular canal has been investigated.

An E-core type of inductive transducer is used to determine the movements of a light oil, which forms a viscous fluid contained in a circular path in a copper tube. The bandwidth of this artificial semicircular canal is 0.02–30 Hz.

When this two-legged device is marching on the spot, it is possible quickly to slide a wooden block beneath one of the feet without the step faltering at all, one of the legs lengthening as the other shortens to compensate for the extremely uneven nature of the newly encountered surface.

A form of leg structure described by Nichols and Witt uses two almost vertical spaced bars for each leg, pivoted separately at the front and the rear of the 'hip' and at the heel and the toe[33]. It was shown that it is advantageous for the heel and toe pivots to be spaced rather more widely than the hip pivots. If the hips of this device are rocked slightly from side to side by a human rider, then the device walks forward. The legs are driven by linkages from the platform on which the rider sits, though it has been suggested that perhaps this linkage might not be absolutely necessary for a human rider. This form of device has been demonstrated in use by a legless person, who was certainly able to get around and was well pleased with his mobility.

Work such as this shows great promise not only for applications to prosthetics but also for application to the robot, and it is hoped that work on the robot will in turn be of help in the prosthetics field. It has been pointed out by Roberts that while in most non-aquatic animals, propulsion and weight support are provided by the same mechanism, it is usual in engineered vehicles to solve the two problems separately[36]. He has shown by construction of a model that the human knee-joint is constructed in such a way that it supports weight quite stably irrespective of the angle at which the knee is set.

Flexible knee-joint

It would be very desirable if a mobile robot could easily climb stairs. Recent work indicates that the mechanical form of a robot leg which can achieve this performance might not be too complex.

Prosthetic legs have been demonstrated which incorporate a free-swinging knee-joint which can be locked against bending. This is done whenever the subject voluntarily flexes the stump muscles in the leg, so causing myo-electric signals to be transmitted to the control system[26]. Although the knee-joint can be locked against bending, it is always free to extend. When the joint is fully extended, the bending-lock magnet is turned off so that the knee-joint can be bent again and the leg can swing.

A human wearing such a prosthetic leg can learn to climb stairs in a few hours of practice. The control system operates from a 6 V battery, which has a life of 24 h of 20 000 flexings.

Thring has publicly demonstrated tracked robot devices which are capable of climbing normal domestic stairs while carrying a considerable load.

Although it would be desirable if the general-purpose robot could climb stairs, the best aim is to endeavour to make the robot so cheap that there will be no difficulty in providing a separate one for each floor.

Ankle-joints

A walking robot might reasonably be expected to move in any environment suitable for a human walking motion. This includes walking on sloping surfaces, and in order to achieve the human degree of flexibility it is desirable that a robot be fitted with a human type of foot mounted on an ankle-joint.

In this connection, the work which has been carried out on the ankle-joint for use in prosthetic legs is of interest. No single medium has been found which can compare with living tissue for durability, fatigue resistance, accommodation and strength-to-weight ratio. However, rubber has been used successfully in the Niedhart suspension for railway systems, and this principle has been adapted for use in the artificial ankle-joint, the work having been done at AWRE.

A rotor shaped like a six-pointed Star of David has been supported in a hollow hexagon by six pieces of rubber cord each about 10 mm diameter. In this way it is possible for the foot to tip forward or backward or sideways about 15° with respect to the leg, and also to twist about 15° to either side.

A truly cost-conscious engineering approach is being used in the development of such ankle-joints.

Large quadruped walking machine[18, 20, 21]

A recently developed walking machine takes the form of a quadruped weighing about 1350 kg. This carries a human operator, and the machine is intended to extend the size and the strength of the operator, primarily for military purposes in difficult terrain conditions. The machine copies the four-legged movements of the human operator. The development of the machine was, in fact, sponsored by the U.S. Army, and it makes use of the force feedback techniques developed for use in manipulators.

It is intended primarily for tasks such as climbing steep slopes, negotiating river banks, etc. This makes the research interesting, though from work such as that of Ballinger[1], the possibilities of the tracked vehicle for such tasks appear as yet to be hardly fully exploited.

The quadruped is 3.3 m high, and it can move at a speed of about 8 km/h on level ground, transporting a 225 kg load. Each of the four legs is 2 m long, and it has two motions in the hip and one in the knee.

Hydraulic actuation is used, the primary power source being a petrol engine. The three motions in each leg are limited in amplitude, and there are no ankle motions or leg-twisting motions. The human leg for comparison has eight distinct motions, added to which are the three possible motion of the pelvis.

Such work is not likely to be of use for incorporation into robot devices for some time, though the detailed lessons learned will be valuable.

The operator sits strapped vertically on a seat, controlling 12 motions simultaneously and responding to 12 force feedback inputs. There is no feedback from the bottom of the feet of the machine. The amplification of the operator's hand and leg motions is 4:1. While the operator was first learning to walk with the machine, it was found that it was not possible to walk successfully if the rear feet of the machine could not be seen by the operator. Consequently, it was found necessary to add rear view mirrors at first, though after some practice the operator was able to adapt to the requirements of the novel mode of locomotion, and was even able to operate the machine when blindfolded.

Various difficulties were encountered by the human in learning to operate the machine. For example, the human is not used to having a horizontal body, and the roll, pitch and yaw movements of the machine are not in phase, while the operator is restricted in that his usual eight degrees of freedom of leg movement are reduced to only three. The dependence of the knee torque on the angle of knee flexing is nonlinear, while the differences in dynamic response, friction, leg forces and so on have all to be adapted to by the operator. One difficulty is that, in walking, the operator is used to having one foot always stationary with respect to the ground, and this is no longer true when he is driving the machine. Because of the restrictions, the machine 'limps'.

The sort of actions that were found to be possible with this quadruped after 15 h of operator training were walking; turning; balancing on three legs (or on two diagonal legs); climbing over 1.2 m obstacles; walking in narrow pathways and weaving between obstacles; lifting one end of a jeep using one foot; pushing a jeep along with the front feet in the vehicle; skidding a 450 kg load along; lifting a 225 kg load with one foot and loading it on to a vehicle; and accurately positioning a load.

The main petrol engine is of 67 kW, and hydraulic pumps, heat exchangers for oil and water with fan cooling, an accumulator, filters, etc., all have to be carried for the hydraulic control system. The hydraulic flow is about 225 litres/min at a pressure of about 15 000 kN/m^2, while a pressure of about 23 000 kN/m^2 is available for maximum force efforts. The control is completely hydraulic, no electronics being used.

It is claimed that equipment like this could be competitive with a helicopter for use on rough terrain, and it has the advantage that the operator does not require a lengthy and expensive training program. The interest of this program from the point of view of robotics is mainly from the demonstration of the capabilities of the quadruped machine. However, in addition, the work does illustrate the difficulties that we can expect in the control of such vehicles with a robot brain mechanism, since the human operator finds that a great deal of co-ordination is required. This takes some time for the human to learn, and we must expect a fairly long training time for robot control of a quadruped vehicle.

Before the complete quadruped vehicle was constructed, some preliminary tests were carried out with a skeletal machine with no power other than that supplied by the operator, and these tests were also successful.

Other anthropomorphous machines[39–52, 70, 71]

Machines such as the quadruped described in the previous section have been called anthropomorphous, in that they resemble man in form and duplicate his manipulative powers. Such work has been largely associated with the name of R. S. Mosher of General Electric. It has been suggested that a shovel can be included in the category of anthropomorphous machines, and it is also claimed that an untrained human is quite capable of using a shovel. Now this is not really true, as witness the formal training which is found to be necessary in some gold mines before workers are regarded as proficient with a shovel. However, there are many advantages in the use of machines to augment the facilities of man, and some of the work is of interest in the field of robotics. For example, it has been pointed out that a machine, unlike man, is quite capable of continuous rotation of a screwdriver.

It is perhaps of interest to mention here an experience of the writer on a flow production line. The foreman of this line was of an inventive turn of mind, and he invented a new screwdriver which was to revolutionise the production. It took the form of a stand, with a handle which was rotated by hand, rather like a pencil-sharpener. The handle rotated a spindle fitted with a screwdriver end. The workpiece into which the screw had to be inserted was held in a clamp; the screw was started by hand; the handle was then turned while sideways pressure was applied; and the screw was rotated and screwed into the workpiece.

Now I found that the invention was most unsuccessful, because there was no 'feel' to it. Every ten screws or so, I simply churned off the screw-head, leaving the shaft of the screw inserted uselessly in the workpiece. I acquired a simple ratchet screwdriver, and found that I had far less trouble and that I could actually double my rate of production. The foreman was annoyed when he found out, but I discovered that I was not the first operator to have abandoned the use of his invention. The point of interest is that the invention was not successful because the operator had little or no 'feel' of the force which he was applying, and no limit such as a slipping clutch.

In the CAMS program, the point has been made that a robot without this sense of 'feel' would wrench doors from their hinges; would pull chairs apart when it lifted them; and would jam rods into pipes if it tried to insert them. Consequently, the use of force feedback has been an important feature of the program.

One product of the program has been the Handyman master–slave manipulator for, for example, nuclear use. This has ten motions in each of two arms, with electric control of hydraulic power. The operator's arms fit into a harness called the 'follower rack'. Tests in carrying out the operation of inserting a peg into a hole using the Handyman showed that there was definite learning with force feedback to the operator, but that the performance was erratic without. The time for the task was reduced to one-half, and the power consumption was reduced to one-third, so that the energy consumption was one-sixth when force feedback was used.

In another experiment in this program, an operator had his head some 4.5 m above the floor, standing on a two-legged structure which he had to balance. The control was by foot flexing and by waist bending. Some operators were understandably so nervous that they were unable to control the balance, and it was found that relaxation was essential.

The Hardiman Exoskeleton was intended to be worn by a man, the intention being to amplify his strength. It was hydraulically controlled at about 20 000 kN/m² and the man wearing it could lift a load of about 700 kg to a height of about 2 m, his strength being magnified some 25 times. Work on the exoskeleton showed that a minimum of five motions was required for each of the legs. In order to achieve balance when a heavy load was being lifted, the length of the feet could be extended. It was regarded as conceivable that a machine of this type could be built with a height of 6 m or even 10 m.

Other work was on the Man-Mate industrial materials handling boom [62, 64, 89]. This was a long arm with a single elbow-joint, the shoulder-joint being at floor level. The jointed arm gave a reach of about 3.5 m maximum and less than 1.5 m minimum, while it could swing in a total arc of 220° from side to side. It could extend at a rate of about 0.45 m/s and hoist at about 0.6 m/s, the rate of swing being 45°/s. Force feedback to the operator was provided in three motions, with a maximum value of about 20 N. The supply was 10 kW, three-phase. A hydraulic pressure of 12 000 kN/m² was used, and the movement locked if the pressure was lost. Vacuum cup, hook or mechanical grippers were available, with a load of about 170 kg or perhaps 270 kg. Man-Mate has been used to handle vitreous enamel dipping of refrigerator shells, conveyor loading and barrel handling.

A reported method of locomotion for lunar vehicles[54] which has been proposed by Professor G. P. Katys of the Russian Institute of Control Problems is, to say the least, somewhat unusual. A long rod is supported by two tripods A and B, the rod overhanging at each end. A heavy weight can slide along the rod from end to end. When it reaches one end, say A, its weight causes the tripod at the other end, B, to be lifted. The rod can then be rotated about the remaining tripod A as centre, and the tripod B can be placed down again at some selected point around the circle of possible movement. The action can then be repeated, but with the weight moved to the other end B of the rod so that the tripod A can be lifted and moved around in a circle.

REFERENCES

1. Ballinger, H. A., 'Machines with Arms', *Sci. J.*, **4**, October, 59 (1968).
2. Johnson, E. G. and Corliss, W. F., *Teleoperators and Human Augmentation*, AEC/NASA Technology Survey, US Govt. Printing Office.
3. Goertz, R. C., 'Fundamentals of General-Purpose Remote Manipulators', *Nucleonics*, **10**, November, 36 (1952).
4. Goertz, R. C. and Thompson, W. M., 'Electronically Controlled Manipulator', *Nucleonics*, **12**, November, 46 (1954).
5. Mosher, R. S., 'An Electrohydraulic Bilateral Servo-manipulator', *Proc. 8th Conf. Hot Labs Equipt, December, 1960* (A.N.S.).
6. Goertz, R. C., 'Manipulator Systems Dvelopment at ANL', *Proc. 12th Conf. Remote Systems Technol., November, 1964* (A.N.S.), 117.
7. Goertz, R. C. and Thompson, W. M., 'Master-Slave Servo-Manipulator Model 2', *Proc. 4th Symp. Hot Labs Equipt, 1955* (O.T.S., Dept of Commerce, Washington), TID 5280.
8. Grace, J. E. A., *et al.*, 'A Radiation Stable Heavy Duty Electromechanical Manipulator', *Proc. 8th Conf. Hot Labs Equipt, December, 1960* (A.N.S.).
9. Goertz, R. C., *et al.*, 'The ANL Model 3 Master-Slave Electric Manipulator', *Proc. 9th Conf. Hot Labs Equipt, November, 1961* (A.N.S.), 121.
10. Young, John F., 'Electronic and Magnetic Amplifier Voltage and Frequency Regulators', *Electricity in Industry*. No. 9 (1956).
11. Young, John F., 'Simplified Feedback Stabilisation', *Process Control Automn*, **10**, June, 233 (1963).
12. Giles, A. F., *Electronic Sensing Devices*, Newnes (1966).
13. Cotes, J. E. and Meade, F., 'The Energy Expenditure and Mechanical Energy Demand in Walking', *Ergonomics*, **3**, 97 (1960).
14. Wilkie, D. R., 'Man as a Source of Mechanical Power', *Ergonomics*, **3**, January, 1 (1960).
15. Frank, A. A., private communication.
16. Vukobratovic, M., Frank, A. A. and Juricic, D., 'On the Stability of Biped Locomotion', *Trans. IEEE*, **BME17**, January, 25 (1970).
17. Frank, A. A. and McGhee, R. B., 'Some Considerations Relating to the Design of Autopilots for Legged Vehicles', *Symp. Aids Hum. Motion*,

January, 1968 (sponsored by General Electric and US Army, Warren, Michigan).
18. Anon., 'The Walking Truck', *Mach. Des.*, **41**, April 17, 32 (1969).
19. Witt, D. C., 'Powered Lower Limb Prosthesis', *Symp. Basic Problems Prehension, Movement Control Artificial Limbs, London, 1968* (Instn Mech. Engrs), paper 4, 18.
20. McGhee, R. B., 'Finite State Control of Quadruped Locomotion', *Simulation*, **9**, September, 135 (1967).
21. McGhee, R. B., 'Some Finite State Aspects of Legged Locomotion', *Math. Biosci.*, **2**, 67 (1968).
22. Ingram, D. J. and Stonehouse, B. H., 'The GEC Power Manipulator', *G.E.C. Jl*, July, 153 (1958).
23. Anon., 'Remote Handling Equipment', *Des. Components Eng.*, October, 489 (1961).
24. Anon., 'The Machine Walks', *New Scient.*, **47**, September 3, 473 (1970).
25. Anon., 'Electro-Hydraulic Control for Artificial Arm', *Electron. & Power*, **16**, May, 186 (1970).
26. Horn, G. W., E.N.A.S.r.L., Electro-Control, private communication.
27. Ring, N. D. and Welbourn, D. B., 'A Self-Adaptive Gripping Device: Its Design and Performance', *Proc. Instn Mech. Engrs*, **183**, pt 3J, November, 45 (1968).
28. Ferrel, W. R. and Sheridan, T. B., 'Supervisory Control of Remote Manipulation', *IEEE Spectrum*, **4**, October, 81 (1967).
29. Thring, M. W., 'A Preliminary Design Study of a Robot to Operate a Large Industrial Store', *Proc. Joint Conf. Electron. Control Mech. Handl., Nottingham, July, 1971* (IERE), 375.
30. Palmer, J., 'Experimental Proprioceptive Grab', *Proc. Joint Conf. Electron. Control Mech. Handl, Nottingham, July, 1971* (IERE), 385.
31. Young, John F., 'A Stock Control Coding Scheme for Industrial Electronics', *Electron. Components*, **9**, March, 295 (1968).
32. Hall, J. I. and Witt, D. C., 'The Development of an Automatically-Stabilised Powered Walking Device', *Proc. Conf. Hum. Locomotor Eng., Sussex, September, 1971* (Instn Mech. Engrs), 131.
33. Nichols, G. K. and Witt, D. C., 'An Experimental Unpowered Walking Aid', *Proc. Conf. Hum. Locomotor Eng., Sussex, September, 1971* (Instn Mech. Engrs), 144.
34. Grieve, D. W., and Cavanagh, P. R., 'Changes of Electromyographic Patterns and Limb Movements Related to the Speed of Walking', *Proc. Conf. Hum. Locomotor Eng., Sussex, September, 1971* (Instn Mech. Engrs), 1.
35. Paul, J. P., 'Comparison of EMG Signals from Leg Muscles with the Corresponding Force Actions', *Proc. Conf. Hum. Locomotor Eng., Sussex, September, 1971* (Instn Mech. Engrs), 13.
36. Roberts, T. D. M., 'Knee-Joint Kinematics and Control-Signal Economy', *Proc. Conf. Hum. Locomotor Eng., Sussex, September, 1971* (Instn Mech. Engrs), 379.
37. McKenzie, D. S., 'Knee Controls for Artificial Legs', in: Kenedi, R. M. (ed.), *Biomechanics and Related Bio-Engineering Topics*, Pergamon (1965).
38. Contini, R., Gage, H. and Drillis, R., 'Human Gait Characteristics', in: Kenedi, R. M. (ed.), *Biomechanics and Related Bio-Engineering Topics*, Pergamon (1965).
39. Mosher, R. S., 'Handyman to Hardiman', *S.A.E. Congress, Detroit, January, 1967*, paper 670088.
40. Mosher, R. S., 'Industrial Manipulators', *Sci. Am.*, October, 88 (1964).

41. Mosher, R. S. and Murphy, W. W., 'Human Control Factors in Walking Machines', *A.S.M.E. Hum. Factors Meet, Chicago, November 7th, 1965*.
42. Mosher, R. S., 'Force Reflection Electrohydraulic Servomanipulator', *Electro-Technology, N.Y.*, (1960).
43. Mosher, R. S., 'Operator-Machine Relationships in the Manipulator', *Electro-Technology, N.Y.*, (1960).
44. Mosher, R. S., 'Exploring the Potential of a Quadruped', *S.A.E. Int. Automotive Eng. Congress, Detroit, January, 1969*, paper 690191.
45. Mosher, R. S., 'Robots to Amplify Man's Manufacturing Efforts' *Proc., A.S.M.E. Des. Eng. Conf., Chicago. 1970*.
46. Barnes, S., 'Machines that Walk', *Mach. Des.*, **37**, February 17, 156 (1965).
47. Barnes, S., 'Army Looks at "Living" Vehicles', *Mach. Des.*, **39**, May 25, 18 (1967).
48. Liston, R. A. and Mosher, R. S., 'A Versatile Walking Truck', *A.S.M.E./N.Y.A.S. Transportation Eng. Conf. Washington, October, 1968*, paper 68 TRAN.
49. Liston, R. A. and Mosher, R. S., 'The Development of a Quadruped Walking Machine', *A.S.M.E., August, 1967*, paper 67TRAN34.
50. Anon., 'Robots Herald New Design Capabilities', *Prod. Eng.*, **41**, February 16, 82 (1970).
51. Berker, M. G., *Theory of Land Locomotion*, University of Michigan Press (1956).
52. Gray, J., *How Animals Move*, Cambridge University Press (1953).
53. Anon., 'Mechanical Arm Can Even Thread Needles', *Electron. Des.*, **19**, November 11, 30 (1971).
54. Anon., 'Leg Up for Luhokhod', *New Scient.*, **53**, January 6, 26 (1972); Katys, G. P., *Aviatsiya Kosmonaut.*, No. 12, 28 (1971).
55. Anon., 'Automated Warehousing—A Slave Computer on Every Stacker Crane', *Mech. Handl.*, **59**, May, 85, (1972).
56. Anon., 'The Humanoids are Coming', *Mech. Eng*, **94**, January, 32 (1972).
57. Fletcher, M. J. and Leonard, F., 'The Principles of Artificial-Hand Design', *Artif. Limbs*, **2**, May, 78 (1955).
58. Godden, A. K., 'Some Factors in the Design of an Adaptive Artificial-Hand', *Proc. Instn Mech. Engrs*, **183**, pt 3J, 50 (1968).
59. Somers, J. C., 'The Manipulator: Its Design and Applications', *Mech. Eng.*, **82**, February, 64 (1960).
60. Harbison, S. G., et al., 'Application of Systems Analysis Techniques to Remote Systems', *Trans. Am. Nucl. Soc.*, **12**, November, 848 (1969).
61. Lilywhite, P. L., 'A Survey of Forging Manipulators and Their Applications', *J. Iron St. Inst.*, **190**, December, 394 (1958).
62. Anon., 'True Arm with Wrist Designed for Handling Materials', *Mach. Des.*, **41**, May 29, 18 (1969).
63. Anon., 'Micromanipulator for Miniature Assemblies', *Engineering, Lond.*, **195**, February 8, 240) 1963).
64. Anon., 'Firm Manipulates Cost Out of Heavy Welding', *Steel*, **153**, August 5, 40 (1963).
65. Sines, G., 'Microclaws for Use in Micromanipulation', *Rev. Sci. Instrum.*, **37**, July 973 (1966).
66. Dolan, C. P., 'High Vacuum Micromanipulation', *Rev. Sci. Instrum.*, **39**, July, 1060 (1968).
67. Anon., 'Manipulators', *Nucl. Eng.*, **7**, 228 (1962).
68. Goertz, R. C., et al., 'The ANL Mk TV2—An Experimental Five-Motion Head-Controlled TV System', *Trans. Am. Nucl. Soc.*, **9**, November, 619 (1966).

69. Anon., 'Rocket Lab Arms Self for Safety', *Chem. Eng.*, **69**, August 6, 72 (1962).
70. Chironis, N. P., 'Can Engineers Soon Make Every Man a Superman?', *Prod. Eng.*, **39**, March 25, 38 (1968).
71. Johsen, E. G., 'Humanoids: The Remote Systems Technology', *Nucl. News*, **12**, March, 37 (1969).
72. Klopsteg, P. E. and Wilson, P. D. (eds), *Human Limbs and Their Substitutes*, McGraw-Hill (1954).
73. Battye, C. K., *et al.*, 'The Use of Myoelectric Currents in the Operation of Prostheses', *J. Bone Jt. Surg.*, **37B**, August, 506 (1955).
74. Anon., 'Brain Waved Artificial Leg', *Sci. Dig.*, **20**, December, 37 (1946).
75. Spielrein, R. E., 'Some Modern Prosthetic and Orthotic Trends and Developments seen as a Challenge to the Engineering Profession', *J. Inst. Eng. Aust.*, **41**, June, 73 (1969).
76. Wedlick, L. T., 'External Power and Recent Concepts in Control of Limb Prostheses', *Med. J. Aust.*, February 8, 278 (1969).
77. Engen, T. J., 'Development toward a Controllable Orthotic System for Restoring Useful Arm and Hand Actions', *Orthop. Prosthet. Appliance J.*, **17**, June, 184 (1963).
78. Weltman, G., 'Myoelectric Control of a High-G Servobrace', *Proc. 5th Int. Conf. Med. Electron.*, *1963*, 627.
79. Anderson, M. H., *Upper Extremity Orthotics*, Thomas (1965).
80. Simson, D. C. and Sunderland, G. D., 'A Position Servo Control System for Powered Prostheses', *Wld Med. Electron. Instrum.*, **3**, May, 116 (1965).
81. Dorcas, D. S. and Scott, R. N., 'A Three-State Myo-electric Control', *Med. Electron. Biol. Eng.*, **4**, July, 367 (1966).
82. McLaurin, C. A., 'The Use of Electricity in Upper Extremity Prostheses', *J. Bone Jt. Surg.*, **47B**, August, 448 (1965).
83. Simpson, D. C., 'Functional Requirements and Systems of Control for Powered Prostheses', *Biomed. Eng.*, **1**, 250 (1966).
84. Scott, R. N., 'Myoelectric Control of Prostheses and Ortheses', *Bull. Prosth. Res.*, **10**–7, Spring, 93 (1967).
85. Baits, J. C., *et al.*, 'The Feasibility of an Adaptive Control Scheme for Artificial Prehension', *Proc. Instn Mech. Engrs*, **183**, pt 3J, 54 (1968).
86. Mann, R. W. and Reimers, S. D., 'Kinesthetic Sensing for the EMG Controlled Boston Arm', *Trans. IEEE*, **MMS11**, March, 110 (1970).
87. Simpson, D. C., 'Control of a Multi-Movement Prosthesis', *Proc. 9th Int. Conf. Med. Biol. Eng.*, *Melbourne, August, 1971*, 230.
88. Newman, N. and Tait, K. E., 'Manipulators: A Survey', *Inst. Engrs Aust., Elect. Eng. Trans.*, **EE8**, April, 1 (1972).
89. George, R. L., 'Materials Handler has Hydraulic Muscle', *Hydrauls Pneum.*, **25**, August, 53 (1972).
90. Hyam, J., 'High Bay but Look No Stacker Cranes', *Mech. Handl.*, **59**, August, 37 (1972).
91. Johnson, E. G. and Corliss, W. R., *Human Applications in Teleoperator Design and Operation*, Wiley (1971).
92. Litvintsev, A. I., 'Vertical Posture Control Mechanisms in Man', *Automn Remote Contr.*, **33**, September, 590 (1972).
93. Okhotsimskii, D. E. and Platonov, A. K., 'Algorithms for Controlling the Motion of a Walking Automaton,' *Proc. 5th All-Union Conf. Control Problems, Moscow, 1971*, 171.
94. Mizen, N. J., 'Machines with Strength', *Science J.*, **4**, October, 50 (1968).

9

Practical robots

Copying robots for industrial use[2, 7, 8, 14, 20, 31, 38, 40, 43, 44, 54, 57, 61, 70, 71]

During the whole of this century there has been a growing interest in industrial automation. If the action of a human being can be reduced to a succession of 'Therbligs' or classified movements[12] by a motion study engineer, then there is no fundamental reason why those stereotyped movements should not be made by a machine rather than by a human operator.

As mentioned elsewhere, the only justification usually accepted for the replacement of the human operator is not humanitarian but purely economic. In such circumstances there is a real need for the reduction of the cost of equipment for automation, and the only way at present that such a cost reduction can be achieved is by the adoption of as much standardisation of equipment as possible. The consequent increase of production quantities can lead to a reduction of unit production cost.

The transfer type of production machine[11, 22, 23, 39], in which fairly standard machine tools are rigidly coupled together by an inflexible conveyor system, is widely used—for example, in the motor car industry. However, with such a system the failure of only one of the machine head stations can bring the whole line to a stop.

A more modern approach is the use of fairly standard machine tools coupled by a fairly flexible system of conveyors[22]. Such a 'link-line' system is reasonably adaptable, but there can be problems. The chief difficulty that the writer has met in his experience of this type of system is the problem of misalignment of the workpiece on the conveyor.

Even slight pressure can cause a misalignment which prevents the workpiece from being properly picked up by the machine feed mechanism. The pressure caused by a microswitch mechanism intended to detect the presence of the workpiece is quite sufficient to dislodge a lightweight piece. The writer has had some success with the use of electronic forms of proximity switches in such cases, since these do not require any contact.

Nevertheless, the problem remains, and correct alignment of the work is often not easy. Even a misalignment of the workpiece in a small fraction of a percentage of the total number of cases is unacceptable in any machine having a very high rate of production, if an attempt is to be made to dispense with the services of a human operator altogether.

With this background, and with some of the problems kept clearly in mind, it is now possible to discuss some of the recent equipments which have appeared for mechanical handling applications such as the loading of machines.

The two best-known types of devices are the Unimate made by Unimation (UNIversal autoMATION) and by GKN, and the Versatran (VERSAtile TRANsfer) made by AMF and by Hawker Siddeley. The arrangement for movement adopted with the Unimate can be compared with that used in the control in direction and in elevation of a turret-mounted gun. The movement of the Versatran on the other hand is based on separate control in three dimensions mutually at right angles.

The Unimate has a longer reach than has the Versatran, the minimum reach of the Unimate being approximately equal to the maximum reach of the Versatran, and it has been suggested that this makes them complementary[19]. The Versatran is more nearly a direct replacement for a human in that it requires approximately the same working area and it has approximately the same reach.

The president of Unimation has been quoted[8] as saying that while there were only 260 devices in use by 1970, he predicted that there would be 5000 by 1974. He gave the figure of 600 000 h of operational experience by mid-1970.

It is perhaps of interest to mention that Thring has referred to these devices as 'senseless' robots[21] in the sense that they do not have any form of 'touch' or of 'vision' feedback control.

These automatic industrial manipulators share with the link-line system the advantage that there can be a considerable storage of workpieces between machines. This facilitates servicing, since most of the production line can continue to be used while one machine is serviced and its storage allowed to build up.

Mechanical hands on machine tools[24]

Many different forms of mechanical hand have been used for feeding machines such as press tools. The work is repetitive. The workpiece is picked up, placed in the machine tool and released; and then the hand moves away to safety and to prepare for the next operation. Some hands are driven by the main drive of the machine tool, though others have an independent drive motor. The former type operates at up to 70 strokes/min with a press capacity of between 20 kN and 1000 kN, and safe operation is possible with a clear space for insertion of about 6.5 cm.

The independent hand must have interlocks so that it controls the operation of the machine tool, ensuring that the hand is clear before a working stroke can commence. Safe speeds of 50 strokes/min have been claimed.

A typical commercial hand can handle work up to about 0.34 kg weight, 18 cm diameter, and square or rectangular shapes up to 230 cm^2, with larger pieces being handled at reduced speeds. A placing accuracy of 0.075 mm can be obtained, in any direction.

Various types of gripper are fitted to these machine tool hands—for example, sucker pads or gripper fingers—and in some cases changing from one form to another takes only a few minutes. For steel or iron, electromagnetic hands have been used.

The suction pad type is operated from a rotary vacuum pump, and easy changeover from suction to pressure can be provided. This changeover is sometimes used to blow oily or sticky workpieces from the suction pads to release them. Typically, the air pressure used is 400–600 kN/m^2 and the consumption is about 0.3m^3 of free air per minute.

Twin pneumatically operated opposed fingers have been fitted to some hands, being typically operated by a wedge moved by a pneumatic cylinder. Electromagnetic pick-up devices have been used, though there can be some trouble here from residual magnetism, which can make release difficult. However, in some cases magnetic 'floaters' have been used to cause mutual repulsion of a number of steel workpieces (e.g. flat laminations), though these appear to have been used mainly to facilitate human handling rather than machine handling.

In some cases it has been found possible to load and unload machine tools simultaneously by the use of two arms which move at the same time as the machine tool is operating. Such an operation would be very dangerous with a human operator, but it is safe with the more rigidly controlled robot loader.

Some forms of hand and arm have been made adaptable so that

they can be moved from one machine tool to another as needed, and in some cases the mountings formerly used by the arm and hand have been used to mount an automatic guard so that a human operator can safely use the machine.

Typical hands or pick-up head jaws which have been used include the standard vice, the hook, the pivot, the chisel, the confined and the Neoprene types, in most cases the standard vice type being suitable. This standard type has two jaws, an upper and a lower, each having two raised vice-like gripping teeth. The hook type is used to pick up parts which have a convenient hole into which the hook can be inserted. The pivot has only a single projecting tooth on each of the upper and lower jaws, and this allows the workpiece to tilt as it is being lifted. The chisel type has a flat chisel-shaped lower jaw, with a two point upper jaw. The chisel blade is used in effect to scrape the lower surface of the work away from the machine as the hand is inserted. The confined type is specially designed for any workpiece which has a vertical edge flange, so that the flange is not crushed or marked. The Neoprene type has the upper jaw covered with Neoprene to prevent damage to the surface of the workpiece. It is seen that with such robot hands it has been usual to use a more or less special hand adapted for the job required rather than attempt to fit a more general-purpose hand.

Some forms of loading and unloading device have been movable from one machine to another as stated above. Others have been movable, sometimes being mounted on a separate frame fitted with wheels, so that they can be wheeled out of the way to permit easy access to the machine—for example, for maintenance. The logical next step in this progression is the fully independent loading and unloading robot which can be used with any machine, and in recent years this type has appeared in different forms.

The Planobot[13]

An early programmed robot arm was the Planobot, which was manufactured by the Planet Corporation of Lansing, Michigan. The hand could be moved in a radius of between about 1 m and 1.8 m and the tilt motion of the arm was up to 60°, so that this robot device could be used for loading and unloading a wide range of machine tools and similar machines.

The hand was fixed at the end of a long straight arm which could be extended, tilted up and down and rotated through 360°, about the vertical axis. In addition, the wrist at the end of the arm could be rotated. A total of 45 different positions of the hand could be stored in the command console in the form of the positions of the settings of 45 groups, each consisting of five potentiometers. In each of these groups four of the potentiometers were used to store the required position of the arm and hand, while the fifth potentiometer was used to set the time constant of a resistor–capacitor circuit controlling the length of the dwell of the stepping switch which moved the control from each set of five potentiometers to the next.

This master stepping switch had six switch-arms or levels and 45 different sequential positions. The error between the required position as set on the potentiometer and the actual position of the arm was amplified and so used to operate relays which control the valve solenoids which determine the motion. Hydraulic motors were used for the rotary motions, while double-acting cylinders were used for ram, tilt and clamp operations.

Rapid swing traversing of the Planobot was at a rate of 90°/s, corresponding to a maximum movement of about 3m/s at maximum radius. When the arm approached within about 4 cm of the final position, the speed of movement was slowed down to about 8 cm/s to limit overshooting. The in-and-out ram movement of the arm operated at a speed of about 30 cm/s.

The Planobot was programmed by setting the four control potentiometers at each position so that the hand reached the required location. The switch was then stepped to the next position and the setting operation carried out for this step.

The time control potentiometer at each position was initially set to the maximum time, so that the initial sequencing operation was slow However, if the movement was found to be satisfactory, then the time control potentiometers could be adjusted to reduce the dwelling time at any required steps, and so to speed the over-all action. It was possible to continue any required motion through two or more steps while changing other motions at each step. In this way a quite smooth motion is obtainable. As an example, the hand can retract until it clears an obstacle, and then continue to retract while it swings sideways after the obstacle is cleared.

The Unimate[17, 52]

The Unimate has a single hydraulically powered arm, fitted with a hand having two pneumatically powered fingers. The fingers can take various forms, the type actually fitted depending on the task to be carried out. The different forms of finger are fairly easily interchangeable, being fixed by four screws of about 0.6 cm. Each of the fingers has a restricted movement of 6° about its pivot, so that the length of finger to be used depends on the size and shape of the workpiece to be handled. Typically, if the distance from the pivot to the fingertip is about 15 cm, then the actual finger length is about 12 cm and the fingertip gap movement is about 3.5 cm. The fingers are operated by a twin linkage, converting from linear to angular motion, from a pneumatic cylinder axial with the arm. The clamping force obtained is a maximum when the fingers are closed, the maximum varying from about 1600 N with an air pressure of about 350 kN/m^3 to about 600 N with an air pressure of about 70 kN/m^3. The actual value of clamping force is directly proportional to the cotangent of the angle between the normal to the finger and the operating linkage, so it decreases as the clamp movement increases. It is also inversely proportional to the distance from the pivot to the fingertip.

The maximum load on the Unimate is about 12 kg, though a load of about 36 kg can be handled at a reduced speed. The nominal positioning accuracy of the load is about 1.25 mm. It should be noted, however, that this is with respect to the Unimate and not necessarily with respect to its surroundings or to the machine being loaded. Although the weight of the machine is about 1600 kg, some of the severe floor vibration conditions which the writer has experienced in industry might cause problems here, and it might be necessary to clamp the loader to the machine being loaded. If this is to be done, then it would probably be better if the loader was actually lighter in weight. The clearance dimensions of the Unimate are about 1.5 m × 1.2 m × 1.4 m high, and the device can be carried by a normal fork-lift truck.

Unimate control system

The hand can be positioned by the machine anywhere in a horizontal area of mean radius about 1.8 m (from the centre of movement) ± about 0.5 m, and in a horizontal rotating traverse arc of a total of 220°. The hand is also movable in the vertical plane over an arc of 220° centred on the wrist. The whole arm can also be rotated in a vertical plane about its centre of rotation, so that the operating position of the hand can be adjusted to a minimum of about 15 cm and a maximum of about 2.4 m from the floor. Thus, in effect, the working volume of the hand takes the shape of a solid-angular segment of a hollow sphere.

The arm can rotate about its centre at a maximum speed of 110°/s and the wrist can also rotate about its centre at the same rate. The wrist can swivel from side to side about its centre through an angle of 180° at the same rate of 110°/s.

The basic arrangement of the control system is as shown in the block diagram of *Figure 9.1*. A Parker Hanifan hydraulic cylinder used for actuation is supplied at about 5200 to 6500 kN/m² from a hydraulic power supply via a servo valve which is controlled by a servo amplifier. The hydraulic power supply uses a gear pump driven by a 7.5 kW three-phase motor. An accumulator of about 11 litres capacity is fitted, in order to maintain pressure when the demand exceeds the pump output, while a low-pressure safety switch prevents operation if the pressure falls below about 3200 kN/m². The hydraulic arrangement and actuators have been described by Szabo[31]. When the machine is being manually set up, the input to the servo amplifier comes from a manual control; but when repetitive operations are being carried out, then the input to the servo amplifier comes instead from the comparator.

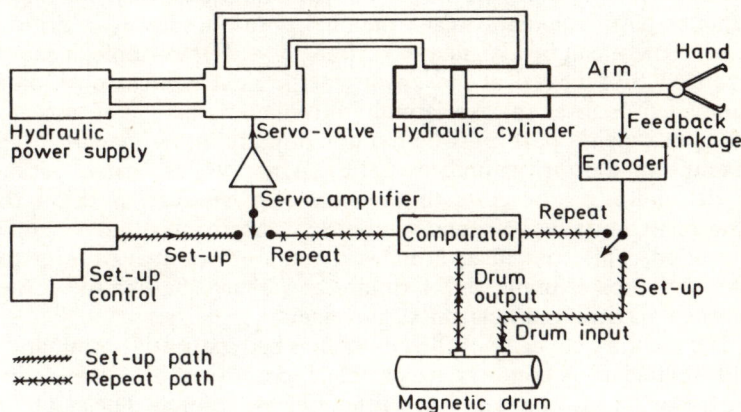

Figure 9.1 Unimate arrangement

Measurements of the actual movements, with respect to the base of the unit, are carried out by an encoder. During the setting-up operation, the digital output from the encoder can be recorded on a magnetic drum, driven by a stepper-motor, in the form of an 80 bit word.

Only the end positions of each required movement are recorded on the drum, and the path taken or the time required are not recorded. An end-of-program signal is used to cause the drum to return to the starting position, and the operation can then be repeated until an external signal causes it to stop.

When the Unimate is in normal operation, with the program set-up on the drum, the output from the encoder is supplied to a digital comparator, where it is compared with the required position as supplied digitally from the drum. When coincidence is achieved in all five of the servo-controlled movements, and when any required external signals are satisfied, then the memory drum is moved on to the next step.

The five degrees of freedom available with this equipment are:

1. Radial extension or retraction of the arm.
2. Vertical depression or elevation of the arm.
3. Rotation to right or left of the arm.
4. Rotation of the hand about the wrist axis.
5. Bending of wrist at right angles to axis.

There is a controlled deceleration as any movement approaches completion, to the point at which numerical coincidence is obtained. All five servo systems normally operate simultaneously, so that a complex path is taken to reach the final position. However, priority of movement can be assigned in various ways. For example, it can be arranged that a vertical movement to the final vertical position is completed before any horizontal movement; this is of use, for example, when a part must be fed horizontally into a machine tool. In addition, a priority number can be assigned to this form of action, so determining how close to the final (e.g. vertical) position the movement must be before any other (e.g. horizontal) movement is permitted. This control determines the amount of curvature in the final path; for example, it determines how sharp shall be the corner between vertical and horizontal movement.

The memory capacity of the Unimate is 200 sequential commands, and isolated 10 A contacts are provided, controlled by the memory, for the switching of any external functions as required. The system is designed for a maximum ambient air temperature of 50°C. The design life of the system is 40 000 h, with a routine maintenance check recommended every 2500 h. The power requirement is 11.5 kVA, three-phase.

It will be seen that this is a quite versatile unit for use when only one, repetitive, sequence of operations is required, with occasional reprogramming, perhaps to a completely different job.

The following figures illustrate the consistency achieved with this device. When used for the feeding of work to a die-casting machine, a Unimate has operated at a rate of 135 pieces/h with a rejection rate of 1 or 2%, compared with the rate of a human operator of 108 pieces/h with a rejection rate of up to 20%.

Some difficulty has been encountered with the Unimate[30] in noisy industrial environments, and it has been found that in some circumstances it was possible for the memory in magnetic cores to be wiped out by extraneous fields. While a paper tape memory does not suffer from this fault, it is not easily alterable. The use of a plated wire form of memory device has proved more suitable for use in noisy environments[27, 46].

Versatran devices[3, 18, 37, 41, 42, 53, 66, 68]

The name Versatran covers a number of different devices produced by AMF and by Hawker Siddeley. The most complex form of device is capable of a wide range of operations. A work load of about 45 kg can be carried by a hand mounted on a wrist which can be rotated about two axes mutually at right angles. The hand is mounted at the end of an arm so that it can be moved horizontally by about ± 40 cm from the mean position and vertically by about ± 40 cm from the mean position, and also rotated about a vertical axis through an arc of $\pm 120°$. The space which can be covered is therefore a segment of a cylinder, about 80 cm in radius and about 80 cm high, the angle of the segment being 240°. The accuracy of positioning is said to be about 1.25 mm, though a figure of ± 3 mm has been given for all axes at maximum reach.

The vertical column of the device is driven by two hydraulic cylinders which rotate the column about its axis by means of chain-and-sprocket drives. Vertical movement of the arm is accomplished by means of a hydraulic cylinder which moves a pinion wheel vertically between two racks, one of which is mounted to the top of the main vertical shaft and the other of which is mounted at its lower end to a massive casting. This casting has a vertical bore surrounding the main vertical column, so that it can be moved vertically by the hydraulic cylinder, rack-and-pinion-arrangement.

The massive casting also has a horizontal bore in which the horizontal arm can travel. This horizontal drive is from a hydraulic motor mounted at the top of the main vertical shaft. The motor rotates a vertical shaft which is rectangular in section, the other end of which is supported in a bearing at the lower end of the main vertical column. A pinion mounted in the main casting slides on the square shaft so that it can be rotated by the shaft while being free to slide up and down with the main casting. The pinion engages with a horizontal rack which is attached to the horizontal arm of the device. Thus as the hydraulic motor at the top of the main shaft is operated, it rotates the vertical rectangular shaft and the pinion, which engages with the horizontal rack and so causes the horizontal arm to move in and out. The wrist and hand are mounted at the end of this horizontal arm.

172 Practical Robots

Two drive rods run through the horizontal arm. One of these rods is rotated by a pinion driven by a rack which is reciprocated by a vertically mounted hydraulic cylinder on the main casting. The other end of this rod rotates the wrist axially to the horizontal arm. The other drive rod runs through the wrist and it is reciprocated by another hydraulic cylinder, this time axial to the arm, which drives a rack backwards and forwards on the hand side of the wrist. This rack engages with a pinion which causes rotation of the hand assembly about an axis at right angles to the arm.

The grippers on the hand are operated by means of a hydraulic supply tube which is flexible and is external to the arm. Various forms of hand can be fitted to carry out special tasks.

The main Brown and Sharpe hydraulic power supply for the unit comes from a motor of about 6 kW which is capable of supplying the required hydraulic fluid at a pressure of about 7000 kN/m² and a rate of about 35 litres/min. In the event of a failure of power, the arm cannot swing around or drop, so ensuring failure to safety by locking valves in the hydraulic system. A 55°C oil thermostat is used.

The general arrangement adopted for control is as that shown in *Figure 9.1*, except that magnetic tape is used rather than a magnetic drum if continuous-path control is required, and potentiometers are used for the memory if a point-to-point control is required.

The unit is about 1.2 m long, about 0.7 m wide and about 1.8 m high over-all, and it weighs about 600 kg. The arm speed is about 0.9 m/s in the vertical or horizontal directions, and the swing rate is 90°/s about the vertical zxis. A load of about 18 kg can be carried at full speed, and the maximum load of about 45 kg can only be carried at reduced speed.

The equipment is air cooled in an ambient air temperature of 0–45°C. The over-all design life, assuming routine maintenance is carried out, is 40 000 h. The electrical power supply required is three-phase, 50 Hz, 6.5 kVA. The equipment can be mounted on small wheels. A longer horizontal arm can be fitted to increase the reach to about 1.4 m.

It has been stated[8] that the downtime on the early Versatran models was as much as 20% but that this has been reduced to 2.5% or less in later models. Such figures are, of course, to be expected with such a new concept. Only the feedback of experience can improve design.

Point-to-point-Versatran

In one form the Versatran is used as a point-to-point controller, where the hand is arranged to move as rapidly as possible between any two points in the working volume. In this form the machine is suitable for such tasks as loading and unloading machines, stacking, etc.

A program drum has been used for storage of the required sequence of operations, though it has also been stated that solid state electronic ring counters made up in modules of ten steps each have been used. The required arm axis positions are stored on a set of potentiometers, 30 potentiometers having been used for each of the three movements. In one form the potentiometers are mounted in modules of six, so that extra modules can be added to expand the memory capacity.

The program unit is mounted in a separate console away from the main robot unit. The program is stored on a removable patchboard, so that different programs can be stored. The required sequence is first laid out graphically on paper and then set up on the patchboard. The sequence is then stepped through manually while the potentiometers are set up for each of the axes to give the required position of the arm and hand.

A typical application for such units has been the automatic loading and unloading of brick-kiln cars. Here it was possible for four Versatran units to replace the labour of 12 men in each of two shifts, and easily to be re-programmed for different sizes of brick. It is of interest to note that standard units were actually mounted upside-down for this application. Bricks are gripped by inflation of gripper bags to 35 kN/m^2.

The maximum number of arm positions in the sequence is typically 30, while the maximum number of steps in the sequence is 100. A total of 12 incoming and outgoing circuits can be controlled by the main unit. On another model, 20 steps are available as standard, though this can be extended, and 18 arm positions are available with the standard set of three 6-potentiometer modules; again this can be extended, by the addition of extra units.

In the operation of loading and unloading, rates between 120 and 1200 per hour are achieved. Forgings have been handled at temperatures up to 1200°C. An application where it has been possible to replace human operators and so to operate in a hazardous environment is the use of the detonation-gun process for the application of coatings. A human could not remain inside the working cubicle in the severe conditions of shock waves and extreme noise level. However, a Versatran unit can be used to handle the loading and unloading of the machines at rates of up to 600 parts/h in this application.

A simplified form of the Versatran is produced, having only two movements. This is known as the Simpltran.

174 Practical Robots

Sub-routine methods have been introduced[31] to facilitate the carrying out of operations which comprise a minor operation repeated over and over, with a small position change taking place between each of these operations. This approach can greatly reduce the number of steps required to be stored in the over-all program. An adaptive gripper control has been introduced to enable the Versatran to pick up individual items from a large group[31, 32]. In some cases the whole device can be mounted on a track up to some 25 m in length, along which it can travel at speeds up to 45 cm/s[31].

The Versatran can be mounted on rails either overhead or on the floor to provide movement from one work station to another[58, 59]. The hydraulic and program-control units do not need to move along with the arm unit. Straight or circular runs of guide rail can be used, and movement at up to about 45 cm/s is possible. Stopping points are pre-programmed, as are arm movements either when stopped or when traversing forwards or backwards. The load capacity of this device is about 68 kg with an arm movement of about 90 cm/s in horizontal and vertical axes. The arm can swing through a horizontal arc of 240° at 90°/s, and the positioning accuracy is about ± 1.5 mm. The workholders at the end of the arm can rotate through 360° and sweep through 280°.

Continuous-path Versatran

The alternative control unit for the Versatran machine is used when it is required that the arm should move through a continuously controlled path, rather than through the shortest point-to-point path. For such applications it is required that the arm should be capable of true continuous controlled movement with both the speed and the acceleration controlled over the whole path.

A separate control console is used, having two magnetic tape decks for the digital memory. Programming is done by leading the arm manually through the required sequence of work movements, using a control joystick box which is plugged into the arm. The twin program tapes are used alternately to avoid interruption, while short programs can be repeated over and over on the tapes. The tapes can also provide signals to external equipment and can be controlled by external equipment. Minor program changes can be made by alteration of a portion of the tape recording. Standard instrumentation-type 6 mm magnetic tape is used, and the normal production cycle is 16 min per tape. One of the tapes is rewound while the other is in control of the machine. A tape speed of about 19 cm/s has been used. An automatic transcribing system can be used for the repetitive

copying of short programs on to the tape.
About 12 sockets are provided for the control of external functions.

Miscellaneous robot devices[70, 74, 77]

Various other robot loading and unloading devices have been produced. Examples are the MINIMAN, made by Foster Fluidics; The Wellman Manipulator used in forging; the developments of Nottingham University, produced by Hawker Siddeley; the Hitachi HIVIP industrial robot device; the Taylor Transiva[31]; and the Erie Autoplace[31].

The Foster MINIMAN has a reach of about 30 cm, a lift of about 7.5 cm and a rotation of 90°. The maximum number of steps in the program is 40, with cycle times of from 2 to 55 s. A most important feature of the Miniman is the provision for feedback from a sensor in the head unit so that an object in an unknown position can be searched for and found. There is one fully hydraulically controlled axis and three pneumatically powered axes, moving between mechanical stops. The Miniman is a low-cost unit, being priced typically at less than £500.

At Nottingham University a team led by Professor Heginbotham has developed the MINITRAN, a simple assembly machine now produced by Hawker Siddeley. They have gone on to work on a programmable two-axis assembly machine which will eventually be programmable by an operator who takes the machine through the required motions.

The Hitachi HIVIP Mk 1 discussed on p. 127 uses a Hitac 7250 digital computer to control two TV cameras and a manipulating hand[35, 62, 63, 64]. One camera looks at an orthographic diagram of the object to be finally assembled, while the other camera looks at the collection of piece-parts to be assembled together. A decision is then made by the computer on the most economical sequence of movements for the manipulating hand of the device to follow in order to assemble the parts. The hand has seven possible movements.

A two-axis parts positioning table has also been developed at Nottingham University, making use of photo-electric detection of a workpiece randomly placed on a table, followed by two-axis movement of the table to bring the part to a given position[9, 10].

The arm used in the Stanford Artificial Intelligence project has been briefly described by Paul[25], and it has been stated that it was designed and built at Rancho Los Amigos hospital near Los Angeles as a prosthetic device for a paralysed human arm[26, 29]. It has six degrees of freedom within an area roughly equal to that covered by a human hand and arm. The hand is of the vice-grip type. Printed circuit motors with harmonic drive gear reductions are used for the drives, the first two joints being rotary and the next telescopic, and the three wrist joints have intersecting axes.

The joint position is determined by a structurally integrated potentiometer at each joint, followed by a 12 bit analogue-to-digital converter. The motors are controlled by a 9 bit digital-to-pulse-width converter, which gives a 15 ms pulse width at maximum output. The sampled data servo system uses the PDP6 control computer and has a sampling rate of 60 samples/s.

The Transiva[31], manufactured by B and R. Taylor, can be fitted with various forms of hand such as vice-grip jaws, internal or external caliper grips, vacuum suction head or magnetic head. The lift is between about 10 and 30 cm vertically and the horizontal reach is between about 10 and 60 cm. The arm traverses through 180°, fixed steps being fitted at 0, 90 and 180°. The wrist can turn through 180°. A load of about 45 kg can be handled at about 1.2 m radius.

The Transiva is sequenced with 16 channels of prepunched holes on a circular drum-type card programmer which trips limit switches controlling the hydraulic solenoid valves. A 1.1 kW electric motor drives the hydraulic pump supplying pressure at about 2800 kN/m².

The Sideman[36], manufactured by Mitsubishi, has been used to load and to unload a sheet metal press at rates up to 940 pieces/h. The power source is compressed air, controlled by an interchangeable plastic rotating preprogrammed drum which carries a number of 12 mm diameter hand-set cams.

Other industrial robots are the Fleximan[50] and the Conslarm[47], made by IHI in Australia. Interest in automatic assembly machines is also increasing[55].

The Erie Autoplace[31] is a low-cost arm, with a 400–600 kN/m² compressed-air control, which can handle loads of between about 28 g and about 4.5 kg. Four different stainless steel arms are available, the arm fitted depending on whether rotation, turnover or reach, or some combination is required. Hands having one fixed and one moving finger, two simultaneously moving gripping fingers, a movable wrist or a vacuum cup are available. Various modes of operation, such as continuous operation or single-cycle operation are possible, and the operations can be interlocked with external equipment. The Autoplace can lift about 7.5 cm vertically, reach or retract through about 30 cm, rotate through 22° and turn over radially through 270°, the maximum reach of the arm being about 60 cm. The motion capabilities of this arm can be described by a volume bounded by concentric spheres[33].

The Norwegian Trallfa robot, by Underhaug and by the DeVilbiss

Company, is intended to apply coatings by mean of a spray gun[75, 76]. A human first takes the spray arm through the required movement sequence, while the outputs of resolvers fitted at the joints are recorded on magnetic tape at some 80 values per second. The robot can then repeat the motions under its own electrohydraulic control in accordance with the recordings on the magnetic tape. The presence of work can be detected photoelectrically, ultrasonically, or by limit switch, and the cycle start can then be delayed by up to 6 s to give the work time to move into position. The tape can store a program up to 109 s long. The maximum speed of the arm is 1.7 m/s and the wrist can swing through 210° horizontally or vertically. The load capacity is about 30 kg at low speed or 15 kg at high speed. The total weight is 650 kg, and the power consumption is 4.5 kW. The arm comprises a vertical part about 75 cm long, which can swing forward and backward by $\pm 80°$, and on which is mounted a horizontal arm about 1.6 cm long. The horizontal arm can swing $\pm 75°$ from side to side and $\pm 68°$ in a vertical plane. The floor-mounted base measures about 1.1 m × 0.75 m × 0.83 m high, so that the end of the vertical part of the arm is about 1.85 m from the floor.

The interaction of the industrial robot, possibly mobile, with its environment, has been considered by Belcher et al.[65]

Comparative economics of robots and humans[72, 73, 78]

Economics of Versatran

It is of interest to give some figures quoted by AMF, the manufacturers of Versatran, for the comparative economics of the machine compared with those of employing a man on the same job. The original figures have been converted to percentages.

Assume that the hourly cost of employing a man to carry out a certain operation is 100 units. This is stated to include wages and fringe benefits, etc., so presumably normal factory overheads are not considered in this study. It is then assumed that the cost of the robot equipment is equivalent to a capital investment of 472 000 units, i.e. 4720 times the hourly cost of human labour. A straight line depreciation over 10 years is assumed. Finally, it is assumed that a single robot can only do the work of one man, both the man and the robot being assumed to work continuously, with one shift working amounting to 2000 h work per year.

The cost of this machine on a 5 year lease is given as 10 920 units/month. Then the hourly cost of operating the robot on a one-shift basis is 80.7 units, of which 65.4 units is the cost of the rental.

On a two-shift basis, the cost is 45.6 units; while if three-shift working is adopted, the cost is down to 32.6 units/h. Notice that this compares with the assumed cost of 100 units/h for human labour.

The cost of the machine if it is purchased outright and assuming a 10 year straight line depreciation to write-off, comes down to 3940 units/month. In this case the hourly cost of operating the robot is 38.9 units, of which 23.6 units is depreciation. On a two-shift basis, the cost is 24.6 units; while on a three-shift basis the cost is down to 18.66 units/h, compared with 100 units assumed for human labour.

The manufacturer of this device also points out that these figures do not include the favourable effect of reduction of the reject rate or the effects on other equipment of the consistent cycling times obtained over long periods.

Szabo has suggested[31], without referring to a specific device, that with an estimated mean time between failures of 600 h, less than 2% downtime and an expected life of 40 000 hours, robots are competitive with human labour. The estimated investment payback period is between 1 and $2\frac{1}{2}$ years, depending on the production rate and the number of shifts worked.

Economics of Unimate[60, 67]

Lindbom has suggested that the criteria to be adopted in selecting an industrial robot application include the following. First, the attributes of the robot, such as speed of movement, weight of the workpiece which can be handled, positioning accuracy, manipulative ability and effectiveness of the robot memory capacity, should be considered. Next, factors concerned with the interfacing between the robot and the process, such as the orientation (or the non-orientation) of the workpiece by the process, the uniformity or non-uniformity of the workpieces, and the difficulty of handling the working material, as is the case, for example, when working with fabrics or with leather, should be taken into account.

Management considerations, such as the acceptance of the value of new methods and of innovations, or the possible existence of other ideas and programmes, are important, and of course the all-important effect on labour relations must be carefully considered.

Economic considerations must loom large in any discussion of the criteria to be adopted when deciding whether or not to introduce industrial robots. Alternative methods, such as special-purpose automation, might be more economically suitable for the particular application being considered. Other economic factors, such as the number of shifts worked, the duration of the job, the cost of the human labour, the cost of installation of the robot and of any auxiliary equipment such as parts, orientation devices, conveyors or inspection devices, can all be very important. Not only the initial cost but also the cost of servicing and of maintenance of the robot must be considered and must be compared with that involved with the equivalent special-purpose automation equipment required for performance of the same job. The overriding economic consideration is, of course, the availability or non-availability of the necessary capital funding for investment in robot equipment.

The payback period P taken to recover the initial cost of a robot is given by division of the total investment I in the robot and its accessories by the difference between the annual saving in labour cost L and the annual cost of maintenance M:

$$P = \frac{I}{L-M} \text{ years}$$

As an example, suppose that the investment I is 25 000 dollars, the labour saving L is 11 000 dollars, and the maintenance cost is 2000 dollars, then the payback period is 2.7 years, assuming one-shift per day. If, on the other hand, two-shift working is adopted, then the labour saving is doubled, though the maintenance cost will only be increased by some 50%, and this change will reduce the payback period to 1.3 years. Lindbom suggests that 3 years is a typical payback period for special-purpose automation equipment, because of its relatively rapid obsolescence, but a realistic figure for robots, as based on over one million operating hours, is some 5 years.

Lindbom further refines his equation by adding in the effect of the factors, capital value Z of the equipment served by the robot, taken on an annual basis and typically some 15% of the cost of acquisition, and a factor q, which is the fractional increase or decrease of the production rate produced by the introduction of a robot instead of a human operator.

The resulting refined equation is:

$$P = \frac{I}{L - M \pm q(L+Z)} \text{ years}$$

Taking the same figures as above, but including a value of q of $+0.2$ (i.e. 20%) and a value of Z of 15% of a capital value of equipment served by the robot of, say, 200 000 dollars, i.e. $Z = 30\,000$ dollars, gives a payback period of P of only 0.85 years. If, on the other hand, it happens that the robot is actually slower in operation than the human replaced, then the negative sign applies in the equation and

180   Practical Robots

with the same figures the payback period becomes $P = 2.9$ years, on the assumption of two-shift working.

If the robot and its accessories cost 25 000 dollars and a straight-line depreciation of 5000 dollars for 5 years is assumed, with a maintenance cost of 75 cents per hour of operation and a human labour cost of $5\frac{1}{2}$ dollars per hour including fringe benefits, then Lindbom gives the return on investment, assuming two-shift operation, of 56%. He suggests that in order to keep the additional overheads down, there should be a potential application for three to five robots in a department before one is purchased since this reduces the cost of, for example, routine maintenance.

Heginbotham[72, 73] has analysed robot economics.

Problems in special environments

In some industrial environments slight leakage of hydraulic fluid produces no problems, and a robot or a manipulator can be regularly hosed down. However, other environments lead to more critical sealing requirements.

For example, the problems introduced by leakage of hydraulic fluid, actual or potential, in a nuclear environment virtually rule out the use of hydraulic controls. Similarly, the leakage of hydraulic fluid is most undesirable in space applications. The possibility of leakage of fluid in a domestic environment is so repugnant to the housewife that hydraulics cannot be used, even though the domestic robot can clean up its own mess.

Pneumatic drives do not suffer from the same limitation, and indeed the use of pneumatic power is not uncommon in radioactive hot cells. The air or gas used can be filtered and cleaned if necessary to avoid contamination. In space it is very necessary to avoid freezing of any fluid, and very careful drying of a fluid would be required at the very least. Probably electrical drives, completely sealed, are to be preferred in space.

Various special manipulators have been produced for underwater use. An early device used a model 300 General Mills manipulator with added sealing, and this was used at depths of up to about 10 m. Pressurised air was pumped into this arm in order to maintain a positive pressure and so prevent leakage of water into the mechanism. Later devices were designed to be resistant to corrosion when the humidity is high.

In some environments the robot control system will be subjected to severe electrical interference or to radiation effects[4]. While shielding is possible in both cases, as also in the less common case of severe

Practical Robots 181

magnetic interference, any such shielding has the unfortunate effect of adding to the over-all weight of the robot. Since the eventual environment of a robot is unknown at the design and manufacturing stage, care must be taken that the control system is as free from the effects of interference as possible.

In some cases it will be necessary to use mobile robots in conditions of severe mechanical shock and vibration. Fortunately, a great deal is known about these problems[15, 16], and so the main problem is likely to be one of achieving satisfactory isolation from shock and vibration at a low cost and without adding excessively to the weight. The former problem is important for the domestic robot, and the latter is important in any robot for mobile applications.

Undersea manipulators[1, 5, 48, 49, 51]

A form of directly controlled manipulator was fitted to early deep-water, high-pressure diving suits. In the 1950s, however, interest in deep-sea operations increased, both from a naval and from a civilian 'farming' viewpoint.

Following on early work with the General Mills model 300 as a swimming pool manipulator, a model 500 manipulator was modified and mounted on a tractor capable of operating at depths of about 1500 m, and this unit was used in the Pacific Ocean.

The Bathyscaph *Trieste* was fitted with a model 150 manipulator for underwater use. This was modified in various ways, being fitted with bellows to compensate for pressure changes as well as being filled with oil. It has not proved to be practicable to seal movable joints effectively against a high pressure differential, and it is consequently necessary to operate underwater equipment at the correct pressure corresponding to the depth at which it is operating.

The problems encountered included those of insulation and of brush wear, while slipping clutches are difficult to produce in a pressurised oil environment. Hydraulic drives present a number of problems for underwater use, including the usual one of leakage, and at the present time electrical drives appear to be preferable.

The use of underwater robot devices for such applications as mining and salvage and for oil drilling and exploration is likely to increase in the next few years. In such devices the effective reduction of weight when submerged is an advantage.

Remotely controlled, towed vehicles fitted with visual, acoustic and magnetic forms of detector have been used for deep underwater work in the ocean[28]. A popular robot used for implanting anchors in deep water is the French device Capsub, made by Techniques Louis

Menard of Paris[45]. The descent below the water of the Capsub is controlled by the injection of compressed air into a large ballast tank at the top of the device, which drives out the water in the tank. In this way a range of from 800 kg of positive buoyancy to 1200 kg of negative buoyancy can be obtained. The injection or expulsion is controlled via a cable by an operator in a boat at the surface. The air volume is automatically maintained constant during descent by admission of extra air. The Capsub can be used to install anchors with capacities from about 5000 to 400 000 kg at depths down to about 200 m.

A Russian cable-controlled sea-bed robot is the Krab[56], which has been used to explore an undersea mountain which rises from a depth of 2100 m, with peaks only 60 m below the water surface. The Krab is fitted with an acoustic depth finder which is used to stop its descent 5 m above the bottom. The lights are then turned on and the built-in TV camera is activated. The Krab is fitted with a manipulator with which it can take samples.

It has been suggested that in 1968 there were some 40 undersea research vessels in use, 33 belonging to the U.S.A., 4 to the Soviet Union, three each to Japan and to France, and one to the U.K.

The American Aluminaut vehicle, which together with another vehicle called Alvin was used in the search for the missing hydrogen bomb off the coast of Spain in 1966, has arms which are 5 m long. For some underwater applications very long arms are necessary, and these introduce problems of control both because of the cramped space available for the operator and because of the inertia of the arm and the viscosity of the water in which it moves. In addition, because the main vehicle is of necessity small, the reaction of the action of the arm and hand and of the weight of the load on the vehicle can produce undesirable movement due to counterthrusts.

In an underwater vessel it is very necessary to make it easy to shed an arm completely in the event of failure, since the safety of the underwater crew must be paramount. The pressure compensation mentioned previously can help to prevent 'spastic' movements of the limbs as the water pressure changes with changes of depth.

The problems involved with direct human control of undersea manipulators, not the least of which is that the operator cannot stay down indefinitely and that raising and lowering can take a long time, have produced an incentive for the development of remotely controlled devices in which the operator, equipped with television, remains at the surface.

A robot device for sampling the sea-bed has been produced by Terresearch[34]. This is a tall cylinder with four fixed legs. It is lowered into position from a survey vessel and then it operates automatically to obtain undisturbed seabed core samples. A heavy weight is raised by compressed air and then allowed to fall on to an anvil, so causing the sampling tube to penetrate the seabed and obtain a 5 cm core sample. The number of blows and the extent of the penetration can be signalled back to the ship via wires, provided the weather permits; but the ship need not remain on a fixed station, since the sinking, the sample-taking and the increase of bouyancy for the return to the surface are all under autonomous control of the sampling robot.

The U.S. Navy has an unmanned submersible shaped like a torpedo and intended for under-ice exploration[69]. The device has a Fibreglass hull some 3 m long and it can operate down to depths of 300 m and at distances of 3 km from the launching hole. It has a speed of 5.6 km/h and an endurance of 10 h.

Surface-operated undersea manipulators[6]

The best-known underwater unit having a surface-based operator is the Curv device of the China Lake U.S. Naval Ordnance Station. This is used mainly for the recovery of weapons after test firings. A total of 37 such recoveries was carried out before the Curv was used to recover the lost hydrogen bomb off the coast of Spain, from a depth of about 750 m.

A commercial French company, the Compagnie Generale pour le Developpement Operationel des Recherches Sousmarines, own a similar craft known as the Telenaute. This is capable of movement in any direction at depths up to about 1000 m. The arm fitted to the Telenaute can handle a load of 50 kg at a distance of 1.1 m. The Telenaute has a very open structure, since there is no need for an unmanned device to have a large and pressurised body.

The presence of oil and gas below the sea-bed gives an incentive for the production of robot-type undersea drilling rigs, especially in view of recent accidents to tower-type manned drilling rigs in bad weather. A completely underwater drilling and exploration facility, with remote supervision or control from the surface or even from the shore, would have great attractions for the oil and mineral companies, and it will almost certainly be produced.

In the meantime, the Shell Mobot device can move to an underwater wellhead and clamp two arms on to it. It can then be moved around, operating tools such as a wrench. The Mobot is tubular in shape and about 7.5 m tall, and its weight submerged is about 1500 kg.

The Unimo is a similar device.

REFERENCES

1. Ballinger, H. A., 'Machines with Arms', *Sci. J.*, **4**, October, 58 (1968).
2. Narraway, R., 'The Anatomy of Industrial Robots', *Prod. Des. Eng.*, March, (1969).
3. Garratt, G., 'Versatran Robots', *Machinery*, **114**, January 22, 133 (1969).
4. Spelman, F. A., 'Electrical Interference in Biomedical Systems', *Trans. IEEE*, **EMC7**, 428 (1965).
5. Anon., 'Submersible Work Boats', *Ocean Ind.*, **3**, No. 2 (1968).
6. Karinen, R. S., 'Land-Based Remote Handling Background of Underwater Handling Equipment', *A.S.M.E. Conf. Underwater Technol.*, May, *1965*, paper 65-UNT-7.
7. Rosenblatt, A., 'Robots are Ready to Grapple with Dirty Jobs in Factories', *Electronics*, **40**, March 20, 165 (1967).
8. Spector, L. F., 'The Robots Are Coming—Or Are They?', *Mach. Des.*, **42**, July 9, 38 (1970).
9. Heginbotham, W. B., 'Research Activities in Mechanical Assembly at Nottingham', *Prod. Engr*, **48**, August, 363 (1969).
10. Heginbotham, W. B., 'Automatic Assembly Tomorrow', *Prod. Engr*, **49**, July, 282 (1970).
11. Hannard, A., 'Transfer Machines and Automation', *A.C.E.C. Rev.*, No. 1, 2 (1960).
12. Carson, G. B. (ed.), *Production Handbook*, Ronald Press (1957).
13. Graham, J. M., 'Programmer for Mechanical Arm', *Control Eng.*, **5**, September, 180 (1958).
14. Mosher, R. S., 'Industrial Manipulators', *Sci. Am.*, **211**, October, 88 (1964).
15. Vigness, I., 'The Fundamental Nature of Shock and Vibration', *Electl Mfg.*, June, 89 (1959).
16. Macduff, J. N. and Curreri, J. R., *Vibration Control*, McGraw-Hill (1958).
17. Engelberger, J. F., 'A Robot Factory Worker', *New Scient.*, **29**, February 3, 270 (1966).
18. Kevern, J., 'Robot Lends Hand in Design of Automatic Machinery', *Prod. Eng.*, **38**, March 13, 105 (1967).
19. Hanify, D. W., 'Reprogrammable Control—Industrial Robots', *Proc. Joint Conf. Electron. Control Mech. Handl.*, Nottingham, *July, 1971* (IERE), 349.
20. Parks, J. R. and Bell, D. A., 'Sensory Devices and Industrial Robots', *Proc. Joint Conf. Electron. Control Mech. Handl.*, Nottingham, *July, 1971* (IERE), 349.
21. Thring, M. W., 'A Preliminary Design Study of a Robot to Operate a Large Industrial Store', *Proc. Joint Conf. Electron. Control Mech. Handl.*, Nottingham, *July, 1971* (IERE), 375.
22. Anon., 'Applications of Work Handling Equipment to Linked Machine Tools and Transfer Machines', part 7 of *Guide to Work Handling Equipment for Machine Tools and Presses*, P.E.R.A. (1961).
23. Holbeche, H. W., 'How Austin Developed Unit Construction Transfer Machines', *Machinist*, **99**, February 4, 51 (1955).
24. Anon., 'Work Loading and Unloading Equipment', part 2 of *Guide to Work Handling Equipment for Machine Tools and Presses*, P.E.R.A. (1958).
25. Paul, R., 'Trajectory Control of a Computer Arm', *Proc. 2nd Int. Joint Conf. Artificial Intelligence, London, September, 1971* (Br. Computer Soc.), 385.
26. Feldman, J., *et al.*, 'The Use of Vision and Manipulation to Solve the "Instant Insanity" Puzzle', *Proc. 2nd Int. Joint Conf. Aritificial Intelligence, London, September, 1971* (Br. Computer Soc.), 359.
27. Anon., 'Plated Wire Finds Spot in Productivity Robot', *Electronics*, **44**, August 30, 25 (1971).

28. Buchanan, C. L., 'Deep Ocean Search by Visual, Acoustic and Magnetic Sensors', *Trans. IEEE*, **AU19**, June, 124 (1971).
29. Kay, A. C., 'Manipulators as Terminal Devices', *Proc. IEEE Int. Comput. Group Conf., Washington, June, 1970*, 290.
30. Balmer, T. R., 'Unimate', *Des. News*, **22**, July 5, 36 (1967).
31. Szabo, M., 'Robots Gear for Action with Fluid Power', *Hydrauls Pneum.*, **24**, September, 77 (1971).
32. Anon., 'Versatran Industrial Robot with Adaptive Control', *Mach. Prod. Eng.*, **119**, November 3, 659 (1971).
33. Stefanides, E. J., 'Patchboard Air Valve Logic Controls Robot Manipulator', *Des. News*, **26**, June 12, 53 (1971).
34. Anon., 'Seabed Sampler Tested', *Oceanology*, **6**, October, 22 (1971).
35. Anon., 'Industrial Robot', *Mach. Des.*, **44**, January 13, 39 (1972).
36. Eshelman, R. H., 'Iron Hands get a Brain Cell', *Automot. Ind.*, **146**, May 1, 47 (1972).
37. Fox, D., 'Five Ways to Auto Load Bore and Face Grinders', *Metalwkng. Prod.*, **116**, February, 72 (1972).
38. Anon., 'The Growing Importance of Industrial Robots', *Mach. Prod. Eng.*, **121**, August 23, 253 (1972).
39. Kay, E., 'Buffer Stocks in Automatic Transfer Lines', *Int. J. Prod. Res.*, **10**, April, 155 (1972).
40. Anon., 'Robots—Plenty of Questions, Few Answers', *Mater. Handl. News*, September, 4 (1972).
41. Anon., 'Robot "Proving Ground" ', *Mech. Eng.*, **94**, January, 36 (1972).
42. Anon., 'Versatran Robot Improves Production of Timer Parts', *Mach. Prod. Eng.*, **120**, May 3, 621 (1972).
43. Hoskins, C. E., 'The Robots are Coming', *Machinery*, **77**, October, 33 (1971).
44. Eshelman, R. H., 'Iron Hands get a Brain Cell', *Automot. Inds.*, **146**, May 1, 47 (1972).
45. Edmiston, K., 'Menard's Method of Seafloor Anchoring', *Undersea Technol.*, **13**, May, 35 (1972).
46. Anon., 'Storing Program Sequence for Industrial Robot', *Control Eng.*, **18**, October, 74 (1971).
47. Newman, N. and Tait, K. E., 'Manipulators: A Survey', *Inst. Engrs, Aust., Elect. Eng. Trans.*, **EE8**, April, 1 (1972).
48. Hunley, W. H. and Houck, W. G., 'Underwater Manipulators', *Mech. Eng.*, **88**, March, 35 (1966).
49. Jones, R. A., 'Manipulator Systems: A Means for Doing Underwater Work', *Naval Eng. J.*, **8**, February, 107 (1969).
50. Anon., 'Portable Manipulator', *Automation*, **9**, August, 91 (1962).
51. Clark, J. W., 'Applying Remote Handling Techniques for Marine Science', *Instrum. Soc. Am. Jl*, **8**, September, 58 (1961).
52. Shosan, S., 'Industrial Robots—Their Potential in Today's Plant Operations', *Plant Eng.*, **23**, July 10, 57 (1969).
53. Sutherland, J. M., 'Industrial Robots and Automation', *Computers Automn*, **19**, October, 36 (1970).
54. Anon., 'Industrial Robot Peens Complex Aircraft Parts', *Control Eng.*, **17**, November, 46 (1970).
55. Wick, C. H., 'Trends in Assembly', *Mfg. Eng. Mgmt*, **69**, August, 26 (1972).
56. Anon., 'Soviets Develop Robot Sea Bottom Vehicle', *Undersea Technol.*, **13**, August, 11 (1972).
57. Szabo, M., 'Industrial Robots Catch Engineer's Interest', *Hydrauls Pneum.*, **25**, August, 28 (1972).

58. Anon., 'And Robot Travels Under Own Steam', *Mech. Handl.*, **59**, August, 51 (1972).
59. Anon., 'Versatran Robot with Programmed Traverse Motion', *Machinery*, **121**, July 19, 85 (1972).
60. Lindbom, T. H., 'Robots: Capabilities and Justification', *Mfg. Eng. Mgmt*, **69**, July, 17 (1972).
61. Anon., 'Are Robots Relevant?', *Mech. Handl.*, **59**, September, 29 (1972).
62. Masakasu, E., *et al.*, 'An Intelligent Robot with Cognition and Decision-Making Ability', *Proc. 2nd Int. Joint Conf. Artificial Intelligence, London, 1971* (Br. Comput. Soc.), 350.
63. Masakasu, E., *et al.*, 'A Prototype Intelligent Robot That Assembles Objects from Plan Drawings', *Trans. IEEE*, **C21**, February, 161 (1972).
64. Kigami, S., 'Robot for Industrial Use and Labour Saving', *J. Inst. Electr. Commun. Engrs Japan*, **54**, November, 1531 (1971).
65. Belcher, J. V., *et al.*, 'Electrical Upgrading of Industrial Robots', *Proc. 19th Conf. Remote Systems Technology, Miami, October, 1971*, 75.
66. MacFee, J., 'The Developing Modular Industrial Robot', *Proc. 19th Conf. Remote Systems Technology, Miami, October, 1971*, 55.
67. Engelberger, J. F., 'Economic and Sociological Impact of Industrial Robots', *Proc. 19th Conf. Remote Systems Technology, Miami, October, 1971*, 47.
68. McAllister, J., 'The Industrial Robot Comes of Age', *Electl Rev.*, Suppl., June 18, 27 (1971).
69. Anon., 'Unmanned Under-Ice Sub Successfully Tested by Navy', *Undersea Technol.*, **13**, November, 39 (1972).
70. Lundstrom, G., *et al.*, *Industrial Robots—A Survey*, International Fluidic Services (1972).
71. Johnson, K. G. and Hanify, D. W., 'The Current Status and Impact of Industrial Robot Technology in the United States', *1st Conf. Industrial Robot Technology, Nottingham, March, 1973*.
72. Heginbotham, W. B., 'The Basic Economics of Industrial Mechanization and Automation', *Int. J. Prod. Eng. Res.*, **11**, 147 (1973).
73. Heginbotham, W. B., Factors Influencing Economic Exploitation of Industrial Automation', *Proc. 1st Conf. Industrial Robot Technology, Nottingham, March, 1973*.
74. Noda, K., and Doi, Y., 'Developments of Industrial Robots in Japan,, *1st Conf. Industrial Robot Technology, Nottingham, March, 1973*.
75. Anon., 'DeVilbiss-Trallfa Robot for Paint Spraying, *Mach. Prodn. Eng.*, **122**, March 7, 315 (1973).
76. Haugan, K. M., 'The De Vilbiss-Trallfa Spray Painting Robot', *Proc. 1st Conf. Industrial Robot Technology, Nottingham, March, 1973*.
77. Anon., 'Retab-Tokyo Keiki Industrial Robots', *Mach. Prodn. Eng.*, **122**, March 14, 358 (1973).
78. Morse, T., Monster or Marvel? The Future Could Well Be Either', *Engineer, Lond.*, **236**, April 5, 36 (1973).

10

Human vision

Development of human vision

In general, devices for visual pattern recognition have in the past been manually set up for correct operation by a human. Experiments on so-called 'learning' devices such as the perceptron have incorporated definite routines by which a human can set up the machine for correct recognition. Although this has often been called 'learning', it is important to note that the operation which has been called 'supervision' or 'teaching' by a human would, in any other application, have been simply referred to as a 'setting-up' operation.

When true learning machines are built, it is important to note that the rate of learning must be expected to be relatively slow. This can best be seen from a consideration of the human performance in the operation of learning to see as an infant, since the method of progression on this process has been quite well studied.

When the child is born, the eyes can move in unison, and an object can be fixated but only momentarily and monocularly[1]. Nearby objects can be fixated at the age of 1 week, but it is several weeks before distant objects can be fixated. The age of the child is about 8 weeks before the eyes can converge so that there can be true binocular vision. However, it is much later, after the age of 6 months that hand and eye co-ordination is sufficient to enable objects to be grasped and picked up. From then on, the co-ordination gradually improves and ever-smaller objects can be picked up.

We known something of the development of human vision from the experiences of people who were born blind and who had their sight restored when they were already adult[2]. From such experiences it is known that the process of learning to see is a protracted and often painful experience for the human. We must not, therefore,

188 *Human Vision*

expect too much of robot learning systems. No doubt we can build them in such a way that they can learn more quickly than can a human, but it must not be expected that the learning process will be rapid or easy if the robot has to learn sufficient to be able to perform complex and almost-human tasks.

Fortunately, it will be possible to use a robot teaching device in order to teach a robot eventually without human intervention, and the learning process will not be slowed down by the necessity for rest or for sleep but only by the necessity for routine maintenance procedures.

There has been some work[19-21] on the direct non-optical stimulation of the human retina, and this it is hoped might lead to visual prosthetics for the blind, though there is obviously much work to be done here. The work might also help in the development of the robot eye.

It has been seen in Chapter 2 that the eyes provide some two million of the three million fibres conveying information to the human brain. Visual information is clearly most important to the human. Up to the present time, the use of visual input to practical robot devices has been unusual. In the future, however, it is likely to be a most important feature of the robot nervous system; and, fortunately, at the present time there are signs that the cost of robot eyes is decreasing.

Information capacity of the human eye[3-5]

It is of interest to consider human visual communication requirements from the point of view of information theory, since this can help to suggest the requirements for a robot visual system.

Experiments carried out by Licklider *et al.*[29] have involved human subjects who were required to repeat verbally, and as quickly as possible, stimuli which were presented visually. It was found that it was possible to achieve information rates of about 30 bits/s. However, if the response required was manual pointing to targets, the rate achieved fell to about 15 bits/s. If the subjects were permitted to respond both manually and vocally simultaneously, the rate achieved became the sum of the individual rates—that is, about 45 bits/s. Pierce and Karlin[3] have obtained similar rates where the task was the reading aloud of lists of common monosyllables. Connected prose could be read faster than random lists of words.

About 1% of the 10^{10} neurons in the human brain are believed to be entirely devoted to the visual process, while as many as 60% are believed to be partially involved in vision. Although there are some 2×10^8 receptor nerves in the retina of the human, these are connected only to about 2×10^6 outgoing nerve fibres. However, it is likely that the number of nerve fibres increases before they reach the human cortex.

Jacobson[4] has estimated that the maximum information transmission capacity of the human eye is about 4.3×10^6 bits/s. This figure was obtained by combining measurements of the flicker fusion frequency with data obtained from experiments on the limits of monocular visual acuity. The average capacity of each optic nerve fibre is therefore about 5 bits/s. Kelley has estimated a figure of about 10^9 bits/s if the level of the luminance is high. This is greater than the information rate obtained from a TV screen, though less than that obtained from a cinema screen.

Human vision is affected by a number of factors such as shape, contrast, brightness, luminance of the background, sharpness of edges and viewing time. The sensitivity of the human eye depends on the colour of the light. The maximum sensitivity is obtained with a blue-green light of wavelength 5100 Å. Colour is only detected by the central most sensitive part of the retina, and so is the detail of shape. The peripheral regions of the retina react only to broad contrast of light and dark. Consequently, the human eye can only detect fine detail near to the centre of the visual region, and this ability, known as acuity, falls off very rapidly as the angle of vision is increased with respect to the most detailed region. In general, the relative acuity can be said to be down to 50% at an angle of only 5° from the centre, but this figure is affected by many other factors also: for example, by the contrast and by the viewing time. The peripheral regions give a maximum response to a shorter wavelength of light than does the central region.

The limited nature of the regions of the retina covered by the colour-sensitive cells has the effect of limiting the extent of the angular limits at which various colours can be seen by a fixed eye, and these limits differ in the horizontal and the vertical planes. The limits can be tabulated as follows:

	White	Yellow	Blue	Red	Green
Horiz.	±90°	±60°	±50°	±30°	±30°
Vert.	±65°	±47°	±40°	±22°	±20°

As in most of the nervous system, the performance of the visual system is affected by a lack of oxygen. This has the effect of reducing the sensitivity of the eye by a large amount, a typical reduction factor being about 10 at a pressure equivalent to that obtained at a height of 6000 m.

190 Human Vision

The smallest increment of luminance that can be detected by the human eye depends on the absolute level of luminance as in the following:

			cd/m²		
Luminance	0.0035	0.035	0.35	3.5	35	100
Increment	6%	3%	1.5%	1%	0.8%	0.8%

The smallest detail that can be detected in a good light of, say, 180 cd/m² subtends about 0.5 minute of arc at the eye.

The distribution of retinal cells in the human eye is as shown in *Figure 10.1*. The colour-sensitive cone cells are confined to the central region.

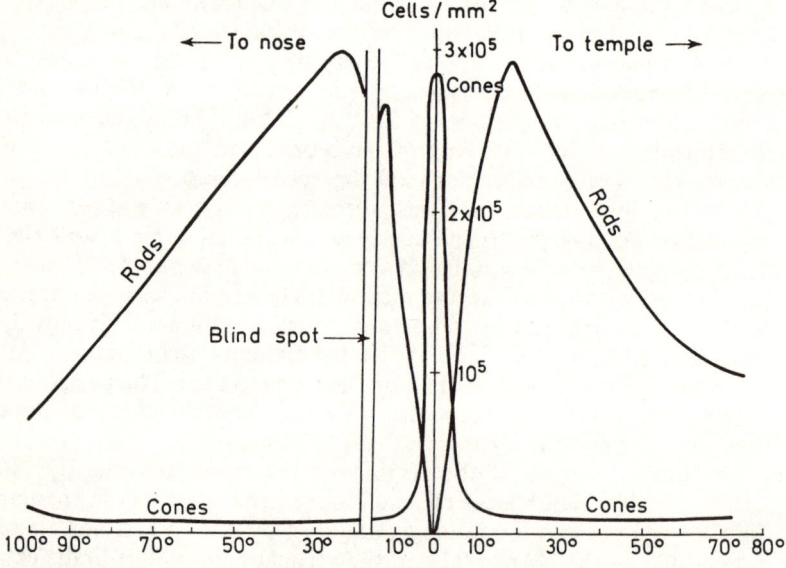

Figure 10.1   Distribution of retinal cells in the human eye

Other features of the eyes

Until more work has been done on the method of acquisition of attention to particular objects from the field of view, it is difficult to envisage the introduction of any means of automatic focusing of the robot eye. The technical problems of the actual focusing are not great. The problem is basically one of a decision on which of the many objects presented to an eye are to be focused on to. Clearly, in this process there must be feedback from the retina to the lens muscles, but the feedback is moderated in some way by the central nervous system.

On the other hand, the servo system controlling the iris diaphragm[7, 9, 27] is a very easy one to implement on a robot. Such devices have been used on cameras. Such a servo would have the effect of reducing the effects of changes of illumination in the scene being viewed. In addition, if the smallest possible aperture is used, then the depth of focusing is greater and there is less need for care in focusing. Indeed, for some initial applications of robots the use of a pin-hole form of lens will be very desirable. Provided that the conditions of illumination are bright enough, the pin-hole lens would eliminate the need for automatic focusing completely.

It is not known, however, how closely the mechanisms of focusing and attention-control are linked, and it might be desirable to retain focusing because of its active link with the mechanism by which the eye and brain concentrate on one object in the field of vision.

One advantage of eliminating any form of iris control in the robot would be that any possibility of oscillation of the controlling servo system would be avoided. It may be that there are other functions of the iris control which have not yet been discovered, since the effective change of area produced (about 16:1) is insufficient to compensate for the brightness range (about 10^5:1) over which the eye is capable of working successfully. These are recent figures as quoted by Gregory[2]. However, it was suggested by Wagman and Nathanson[6] in 1940 that the change of pupil area is only about 5:1, with light intensity variations of at least 10^8. It was also suggested that the area is linearly related to the logarithm of the light intensity over a range of about 10^6, i.e. that the Weber–Fechner law is obeyed in this central region of intensities from about 35 cd/m² down to about 3.5×10^{-5} cd/m².

In the work being carried out at Aston and elsewhere on robot eyes it is very unlikely that the retinas used will ever simulate the human retina directly. This is because, no doubt for evolutionary reasons, the human retina is of a very poor engineering design. The sensitive photo-electric cells of the retina are at the back of the eye. In front of these cells, and between them and the light entering the eye, are various nerve cells and the nerve fibres which carry away the information to the brain. The fibres have to be brought out of the back of the eye somewhere so that they can be taken to the brain, and this is done at a single point. Since there can be no sensitive photo-cells at the point where the fibres are brought from the inside of the eye to the outside, the point forms a blind spot in the eye. All of the light reaching the sensitive cells has to travel through the nerve fibres and

Human Vision

the nerve cells before it can reach the sensitive retina, and it is, of course, seriously attenuated on the way.

Robot eyes will not suffer from this human deficiency. The light-sensitive cells will be facing the light and there will be no need for a blind spot.

One form of automatic iris control has been used on the experimental Picturephone system in which television is combined with the normal speech communication of the telephone. The iris in this system adjusts the lens opening for the variations in ambient light conditions. The iris is controlled by peak averaging of the video signal coming from the camera tube. The iris is fully opened when the ambient luminance falls to about 40 cd/m² and an automatic gain control circuit in the video amplifier then takes over, though this introduces an increasing noise level as the light level is decreased. An advantage of the incorporation of an iris control is that the depth of focus is increased when the illumination level is high and the iris is closed down to a small aperture.

The transfer function of the human iris control loop has been determined by experiments such as that in which oscillation is caused by an increase in the effective gain of the system by illumination of the edge of the iris aperture only by a very intense small spot of light. Such experiments show that the transfer function can be expressed as

$$\exp(-sT_d) \times \frac{G}{(1+sT)^3} \times A \times k \log I$$

where T_d is a time delay in the nerves; T is a time constant; A expresses the adaptation or phase-advancing properties; and $k \log I$ expresses the logarithmic sensitivity to the light flux I falling on to the retina.

The loop is normally stable, with a loop gain of 0.15 and a loop time delay of 0.18 s.

Clynes[8] has given a complex form of possible pupil response, and he has compared actual performance with the predictions using his expressions. While these are of biological interest, it is unlikely to be worth attempting to reproduce the biological response exactly in any robot system, and the complex form will not be further considered.

Binocular vision

Most animals have two eyes. As is well known, the use of two eyes not only has the effect of increasing redundancy, so that a deficiency, even amounting to loss of vision, in one eye is less important.

Although this redundancy is an important engineering feature leading to an increased reliability of the organism, an equally important function of the presence of two eyes is the introduction of distance perception. Objects up to about 6 m from the eyes present sufficiently different images on the two retinas for them to be compared in the nervous system and used to indicate the distance. If the two eyes are both focused on to a given object, then the information about the relative alignment of the eyes can be used to give an indication of the distance of the object.

Except for special-purpose robots, it is unlikely that distance-perceiving binocular vision will be very necessary in the near future. However, the importance of the visual system in the perception of environmental conditions carries the implication that the use of two eyes on a robot is very important from a redundancy point of view. In our work on simple mobile robots at Aston, we have found it useful to fit two simple single-cell 'eyes', each having a different sensitivity. This avoids the necessity for the use of an iris system in such simple eyes, and it has not been used for binocular vision. Such an approach is more akin to the use of rod and cone cells having different sensitivities in the human eye.

There appears to be a form of cross-coupling between the system which rotates the human eyeball and so converges the eyes for viewing close objects, and the higher process of accomodation which seems to be operative at a higher point in the nervous system. A similar cross-coupling effect is required to make the pupil constrict for near vision.

Binocular vision can be used to control a head- or body-rotation control system to centralise the viewed image. However, it is possible to achieve such centralisation with only one eye; in fact, the single-eyed human has to do this continuously.

The robot can if required be fitted with more than two eyes, and there is no need for these to be in the head of the robot. For some applications of robots it will be an advantage if the eye or eyes are fitted elsewhere—for example, in the hand[30].

In experiments with master–slave manipulators fitted with television viewing devices, it has been found that stereo, or 3-D, viewing is a little superior to normal two-dimensional viewing, but only if the three-dimensional television is correctly aligned. If the alignment is not good, then two-dimensional viewing leads to a better performance[10].

A fluidic form of photo-detector has been developed[11]. This gives a signal which depends on the alignment of the detector with respect to the light source, the sun having been proposed for the latter in a space vehicle attitude control.

It has been suggested[12] that binocular vision has the advantage that there is less degredation of a pattern caused by visual noise if vision is binocular rather than monocular. It is an interesting fact that it can take some time for the human brain to perceive the basic pattern in any puzzling picture. This is particularly noticeable in the case of certain stereograms such as those produced by Julesz[13, 14].

Various forms of direction-sensing device using electro-optics have been developed[25, 26].

Colour sensitivity

The cone cells in the human eye are sensitive to colour. It is quite possible to build colour-sensitive photo-electric equipment. A typical example in the writer's experience of a robot-type device which was built commercially using colour-sensitive cells is a device for sorting coloured sweets, to ensure that a selection is inserted into each packet. (Perhaps it should be mentioned that such equipment is necessary when previously filled packets have broken open and there is consequently a mixture of the coloured sweets.) Particular coloured filters are placed in front of photo-electric cells in order to permit only one colour of light to reach any particular cell.

Colour television makes use of filters in front of the camera tubes in a similar way. Thus there is no basic difficulty in the use of colour-detection devices in robots. However, such devices inevitably add to the initial cost of the equipment. In addition, it must be remembered that many animals, and in fact 10% of men, are incapable of detecting colour differences: they are colour-blind. Since the animals and the people manage to survive without colour vision, only those robots which have to carry out tasks involving colour sensitivity need to be fitted with such special apparatus.

However, the use of both rods and cones in robot retinas, to extend the sensitivity range though not the colour range, is probably a desirable thing, even in quite simple robots.

While the sensitivity of the human eye is limited to the normal visible spectrum of colours, there is no reason at all why the robot eye should be similarly limited. While it is probably not worth while making a general-purpose or domestic robot with exceptional colour sensitivity, special-purpose robots could easily be constructed with a visual system which is sensitive to infra-red or to ultra-violet, to radio microwaves or to X-rays, or indeed to any form of electromagnetic radiation for which sensing methods are known. Such special-purpose robots will be extremely useful for some tasks. As an example, an infra-red-sensitive mobile robot would be capable of detecting such faults as hot axle boxes on vehicles. As a useful tool to an art dealer, an ultra-violet-sensitive robot will be invaluable. There are many such examples where extreme radiation sensitivity will make the robot a useful tool.

While vision capability in, for example, the infra-red region[15, 18] could be used to make it possible for a robot to see in an environment which appeared dark to a human, unless there is a special reason such as secrecy, it would be much simpler and cheaper merely to provide the robot with headlamps. For some applications it is possible to attach an illuminating lamp directly to the visual system in a rigid manner, so that illumination is always available. Since the lamp will produce some heat, it is advisable that it be isolated from the visual sensor by some poor conductor of heat, such as a thin layer of air. Sensors of this form designed by the writer have been used successfully on industrial equipment for many years[22].

There are various problems associated with underwater vision, and although conventional techniques have been used in the past, it is likely that in order further to improve the performance compared with human vision future work will include techniques such as the use of reflected acoustic pulses for vision and the use of gated laser pulses for illumination.

It is very noticeable that a colour picture which is so coarsely quantised in squares that it makes no sense when properly focused, can become an easily recognisable picture when it is defocused—for example, if a person with defective vision simply removes his spectacles[23].

In the same way that single-line photodetectors have been produced on integrated circuits using a line of photo-cells with integral scanning, so infra-red single-line scanners have been produced. A typical form uses crystals of triglycine sulphate as the infra-red detectors, with amplification carried out by blocks of eight integrated circuit pre-amplifiers using metal-oxide semiconductor field effect transistors[24].

Various forms of flame-sensing equipment have been produced commercially. One type is required to observe over a wide field of view[28]. Some are used for flame and fire protection, to operate extinguishing equipment. Another form is the flame-failure sensor, which is used to prevent the flooding of furnace equipment with unignited fuel oil. Some flame-failure equipment operates on the flicker of the flame, and this ensures fuel cut-off even if the inside of the furnace continues to glow or to give off non-flickering visible or infra-red radiation after the flame has failed.

Optical illusions

In order to construct visual systems having human characteristics intended for installation in robots, it is necessary for us to know as much as possible about the human visual system. In studying an unknown system it is sometimes its pecularities and its deficiencies which can give us a clue to the basic method of operation.

Some of the basic deficiencies of the human optical equipment can be grouped together under the heading of optical illusions. It is very tempting to study such illusions in order to find generalisations which can be applied to the human visual process. To the normal human being optical illusions appear to be a basic universal characteristic which is possessed by all humans. It should be possible to test theories of vision and proposed robot methods of vision by examining their performance when subjected to the sort of visual image which gives rise to illusions in the human.

Unfortunately, these illusions do not seem to be an invariant experience of the whole human race. It has been found that the Zulu, in particular, does not experience many of the illusions experienced by Western man, and it has been suggested that this is because straight lines rarely appear in his cultural surroundings at all. Instead, everything that he sees is curved. On such evidence, it appears that optical illusions are an acquired experience; that they are not an invariant in man; and that therefore they are of little use in cybernetic studies of optical systems, at least at the basic level.

Recent work makes it appear that the whole development of the brain structure, at least in the visual regions, depends almost entirely on the early environment rather than on hereditary factors. For example, Blakemore and Cooper[16] have reared kittens in a dark environment. For 5 h each day, however, they were put into a plastic cylinder, the inside of which was covered with black and white stripes. A cat reared in this way in an environment of vertical stripes seemed to be later almost incapable of seeing a horizontal black or white rod, while a cat reared in a horizontal environment could not see a vertical rod.

Moreover, tests on the actual neurons in the brains of these animals indicated that the actual physical response depended strikingly on the orientation of a bright slit producing the light on the eye. The vertically experienced cats displayed on optimum response to a vertical slit in most of the neurons tested. Horizontally experienced cats, on the other hand, gave a maximum neuronal response to a horizontally oriented slit. The experimenters suggest that perhaps the nervous system adapts to match the probability of occurrence of features in its visual input.

Work such as this is most significant to the cybernetic engineer, since it indicates that the central learning mechanisms of a robot should be extremely generalised and the specialisation of function should be minimised. At Aston we are basing this generalisation on Pavlovian association.

Cleanliness of robot eyes

The full purpose of the blinking of human eyes is not known. Although the main purpose does not seem to be the maintenance of cleanliness, this is certainly one of the functions. The presence of moisture and of tears also helps in this respect.

It has been unusual in the past to fit special cleaning equipment to engineering photo-electric equipment, although dirt in the optical system has certainly been a cause of failure in some cases. Instead, there has sometimes been a policy of routine maintenance in which the optical system is cleaned and the light sources and the photo-electric cells are replaced at regular intervals.

While it is not impossible to fit devices such as screen wipers to photo-electric optical systems, it is unlikely that this practice will become widespread in robot applications of the near future. In particular, with mobile robot devices routine maintenance will be very important in order to ensure safety, and the eye equipment will have to be maintained at the same time as the rest of the robot.

However, the development of reliable self-cleaning optical systems for photo-electric equipment is likely to be one of the beneficial effects that work on the development of the mobile robot will generate. In some applications of robots, cooling equipment or anti-steam equipment will be required for the eyes.

Eyelids which can be closed are a useful feature in the human from several viewpoints. For example, they make it possible to cut off the visual scene—for example, for sleep or for concentration. In the robot, however, a similar cut-off can be obtained by electrical switching without mechanical movement. As an additional protective reflex to the action of the pupil in cutting off bright light, and also as a protective against mechanical damage, blinking is again a useful feature in the human. A rapid response shutter which could rapidly cut off light would certainly be a useful feature if it could be made simple. A notable example is the failure of the TV camera during the recent human landing on the moon. This could have been avoided had some simple automatic shutter been fitted. There is here a possibility of using the properties of various suspensions in liquids to achieve such an automatic protection rapidly.

There can be no doubt that, for some tasks, the need of the human to blink, and the psychological nature of the drives to blinking can be a big disadvantage. In some cases it is found that the human vision can be obscured for as much as 20% of the time because of blinking[17]. This blank period can be reduced or eliminated completely in a robot visual observer, though, as mentioned before, this will necessitate loss of the mechanical protection afforded by blinking. A cleaning action can be obtained with a much reduced loss of vision compared with blinking.

REFERENCES

1. Gesell, A., 'Infant Vision', *Sci. Am.*, **182**, February, 20 (1950).
2. Gregory, R. L., *Eye and Brain*, Weidenfeld and Nicolson (1967).
3. Pierce, J. R. and Karlin, J. E., 'Information Rate of a Human Channel', *Proc. IRE*, **45**, March, 368 (1967).
4. Jacobson, H., 'The Informational Capacity of the Human Eye', *Science, N.Y.*, **113**, 292 (1951).
5. Carne, E. B., *Artificial Intelligence Techniques*, Macmillan (1965).
6. Wagman, I. H. and Nathanson, L. M., *Proc. Soc. Exp. Biol. Med.*, **49**, 466 (1942).
7. Stark, L., 'Stability, Oscillations and Noise in the Human Pupil Servo Mechanism', *Proc. IRE*, **47**, November, 11 (1959).
8. Clynes, M., 'Non-Linear Biological Dynamics', *Ann. N.Y. Acad. Sci.*, **98**, October 30, art 4 (1962).
9. Jones, R. W., 'Some Properties of Physiological Regulators', in: Coales, J. F. (ed.), *Automatic and Remote Control*, Vol. 2, 674, Butterworths (1961).
10. Goertz, R., 'Manipulator Systems Development at A.N.L.', *Proc. 12th Conf. Remote Systems Technol., November, 1964* (A.N.S.), 117.
11. Miller, W. V., 'Fluidic Sun Sensor for Solar Pointing Fluidic Attitude Control', *Proc. Joint Autom. Contr. Conf. A.A.C.C., Colarado, 1969*, 545.
12. Giarretto, H., 'The Effects of Stereoscopy on the Recognition of Patterns in Visual Noise', *Hum. Factors*, **10**, 513 (1968).
13. Julesz, B., 'Binocular Depth Perception of Computer Generated Patterns', *B.S.T.Jl*, **39**, 1125 (1960).
14. Julesz, B., 'Local and Global Processes in Visual Perception', *Proc. IEE/NPL, Conf. Pattern Recog., 1968*.
15. Anon, 'Invisible Searchlight for Military TV', *Des. Electron.*, **8**, April/May, 4 (1971).
16. Blakemore, C. and Cooper, G. F., 'Development of the Brain Depends on the Visual Environment', *Nature, Lond.*, **228**, 477 (1970).
17. Lawson, R. W., 'Blinking, Its Role in Physical Measurements', *Nature, Lond.*, **161**, 154 (1948).
18. Morten, F. D., 'Infrared Detectors and their Applications', *Mullard Tech. Comm.*, **10**, May, 75 (1968).
19. Brindley, G. S. and Lewin, W. S., 'The Sensations Produced by Electrical Stimulation of the Visual Cortex', *J. Physiol.*, **196**, 479 (1968).
20. Lin, W. C., *et al.*, 'Feasibility Study of Electronic Multi-Electrode Stimulation System for Visual Cortex Stimulation', *Proc. 23rd Conf. Eng. Med. Biol.*, Vol. 12 (1970).
21. Anon., 'Artificial Vision', *Wireless Wld*, **77**, May, 214 (1971).

22. Mawer, J. C., 'Electronic Colour Registration in Printing', *Ind. Electron.*, **1**, August, 561 (1963).
23. Anon., 'Mechanical Eye Views World in Matrix Form', *Sci. J.*, **4**, July, 17 (1968).
24. Blackburn, H., *et al.*, 'Pyroelectric Detector Arrays for Thermal Imaging', *JIERE*, **42**, August, 369 (1972).
25. Johnston, A. R., 'An Electro-Optic Direction Sensor', *J. Spacecraft Rockets*, **9**, September, 690 (1972).
26. Miller, B., 'Eye-Like Sensor Being Developed', *Aviat. Wk.* 97m October 2, 49 (1972).
27. Anon., 'Silicon Intensifier Target Tube gives Better Undersea View', *Electronics*, **45**, October 9, 47 (1972).
28. Erickson, C. W., 'A Flame Sensor with Uniform Sensitivity over a Large Field of View', *Trans. IEEE*, **ED19**, November, 1178 (1972).
29. Licklider, J. C. R., *et al.*, 'Studies in Speech Hearing and Communication', Final Report Contact W191222014, M.I.T. Acoustic Laboratory (September, 1954).
30. Heginbotham, W.B., *et al.*, 'The Nottingham SIRCH Assembly Robot,' *Proc. 1st Conf. Industrial Robot Technology, Nottingham, March, 1973*, 129.

11

Character recognition

Visual character recognition

A vast amount of work has been done in recent years on various methods of machine recognition of visual printed characters. The writer has reviewed such work elsewhere[3].

The recognition by a machine of characters when they are printed on a cheque in a very special script which is designed to be also readable by a human being represents a good technical achievement. Unfortunately it would also seem to be a quite unnecessary waste[11]. If the number to be recognised was printed both in, say, Morse code for the computer and in typescript for the human, nothing would be lost at all, since there is plenty of room on the face of a cheque for both sets of characters. Even for use on more compact documents it should not be beyond human ingenuity to devise a computer code using, for example, a maximum of four dots, which could be used to underline each number printed simply in a normal human typescript*.

With a maximum of six underlining dots a whole alphanumeric alphabet set could be accomodated. There would seem to be little need at all for the complicated machinery which has had to be developed and designed to be capable of reading a special, almost-human, script. There are objections such as noise sensitivity to a simple dot system, but such problems are not insuperable with automatic error detection systems.

Of course, as an engineer one must hesitate to criticise spending on development, since this much-needed money might help to finance future developments in this field.

No attempt will be made here to review methods which have been used for character recognition in computers, since an excellent review has been prepared by the British Computer Society[2]. Instead

* Such a system has now been introduced by Dataflow.[1]

more general work on the simulation of animal visual systems for robot use will be considered. A list of commercially available equipments has been published elsewhere[13]. The field was surveyed by Wee[16], and the economics were discussed by Carsbury[18].

Perceptrons

A great deal of money appears to have been spent all over the world on very expensive experiments in which simple majority logic units have been set up ('trained') to dichotomise, i.e. to differentiate between two different arrangements of inputs.

It is perhaps worth mentioning that once a Perceptron has been set up to differentiate, say, A from B, if it is then required to reset it in order to make it differentiate A from C, it is necessary to re-present A in the course of the setting-up operation. In other words, a Perceptron does not acquire any inherent method of detecting the property of 'A-ness' from its setting-up operation. Instead it is merely set up to differentiate between the two specific patterns. As seen elsewhere[12], the simple demonstration Perceptron unit described can only categorise an F or an L as though it were an X, the important point being that all three letters can be regarded as being equally different from Y (or from O).

However, with greater complication it has been possible to differentiate between greater numbers of different patterns. Such complication does not appear to have been carried very far in practice, perhaps because it involves great expense.

There has thus been Perceptron work using a random form of cross-connection between the S units and the A units, in addition to the basic extensive work using majority logic. There has also been work using the same approach under the different name of Adaline, and a collection of many Adalines with a common output, which has been called a Madaline.

The same approach is used in the so-called Bayes net, except that here the weights given to the inputs by setting up the resistors are usually arbitrarily chosen to represent the logarithm of the probability of occurrence of the various inputs.

To return to the extensive work on the Perceptron type of device; Nievergelt has pointed out that simple geometric properties such as connectedness cannot be recognised by such equipment, and the Perceptron type of device which can recognise a triangle or a circle turns out be be very big[4]. However, although the Perceptron type of approach, under its many different names, appears to be leading nowhere, it does at least use majority logic, as opposed to the binary form of logic. The animal nervous system is not binary in nature or operation.

Other Perceptron work

In some Perceptrons, such as that described by Hay *et al.*[5], or in Taylor's very extensive version in London[6, 7], the variable resistors have been motorised to give an automatic adjustment of value depending on the output of the amplifier.

Taylor's version has been used to differentiate between ten visual patterns such as photographs of faces. Holmes has detected aircraft and buildings[14].

Financial support is still forthcoming for Perceptron work. For example, in this country it appears that attempts are being made to miniaturise Perceptron units by use of integrated circuit techniques.

The shadow mask technique

In the shadow mask technique sometimes used for optical character recognition, a positive version of the image to be recognised is covered in turn by a positive mask of each of the possible images recognised by the system. Any area of the figure being tested which lies outside each of the positive master tasks is then summated individually. A negative version of the image to be recognised is similarly compared with all possible negative images, and in each case the area lying outside each of the master masks is summated. The positive and negative sums are added. The decision is then made that the character being presented is probably that character which gives the smallest over-all sum, sometimes called the 'error-sum'.

In the Cybernetics Laboratory at Aston this method has been investigated by K. P. Fisher[17], using a digital computer. He calls this the method of internal inhibition. To illustrate the method used, consider a simple retina composed only of nine cells. If this retina is presented with the formalised characters '1' and 'I', then the resulting patterns of stimulation would be as follows.

+ve			−ve			+ve			−ve		
0	1	0	−1	0	−1	1	1	1	0	0	0
0	1	0	−1	0	−1	0	1	0	−1	0	−1
0	1	0	−1	0	−1	1	1	1	0	0	0

These patterns of stimulation can be regarded as positive and negative stored masks. Now if any unknown pattern of stimulation is presented, it is compared with the positive masks and its inverse is compared with all negative masks, and any agreements between the presented information and the information stored in the memory is scored one point, while disagreements are ignored.

Thus consider a character '1' to be applied. the corresponding stimulation patterns from the retina are

$$\begin{array}{ccc} 0 & 1 & 0 \\ 0 & 1 & 0 \\ 0 & 1 & 0 \end{array} \qquad \begin{array}{ccc} -1 & 0 & -1 \\ -1 & 0 & -1 \\ -1 & 0 & -1 \end{array}$$

When this pattern is applied, the individual cell stimuli are compared with the information in the store about possible patterns. If these incident patterns are compared in this way with the stored information about the patterns '1' and 'I', then the following results are obtained:

	Compared with '1'	Compared with 'I'
+ve correlations	3	3
−ve correlations	6	2
Sum	9	5

Thus, if these sums are used as a measure of the correlations, it is almost twice as certain that this presented character is a '1' than that it is an 'I'. Note that a perfect correlation with the previously stored pattern is indicated, as in the case given here, by a correlation value equal to the number of cells, nine in this case.

Methods such as those discussed here are sometimes extended for use in noise reduction with retinas containing a great many elements. For example[15], if fewer than 5 out of any square of 9 cells in the array are coloured black, then that particular square is ignored.

Other inhibitory forms

Other methods of using inhibition in a correlating technique can be used instead of the shadow mask (or internal inhibition) method, which involves both the storage and the application of separate positive and negative matrices.

For example, the stored patterns can be the same as with the previously described method, but the testing input obtained from any applied pattern can contain both positive (dark) and negative

204 Character Recognition

(light) signals. To illustrate this, of a '1' is applied, the resulting signals from the matrix of cells are

$$\begin{array}{ccc} -1 & +1 & -1 \\ -1 & +1 & -1 \\ -1 & +1 & -1 \end{array}$$

The resulting positive and negative correlations are then given by

	'1' applied	'I' applied
+ve correlations	+3	−1
−ve correlations	+6	+2
Sum	+9	+1

It is seen that '1' is now differentiated from 'I' by 89%. Similarly, 'I' is differentiated from '1' by 89%.

A further possible method involves the storage in the memory only of the positive associations:

```
   '1'          'I'
 0 1 0        1 1 1
 0 1 0        0 1 0
 0 1 0        1 1 1
```

However, the testing input obtained from any applied pattern can again contain both positive (dark) and negative (light) signals. Consequently, as in the previous case above, if a '1' is applied then the resulting signals from the matrix of cells are

$$\begin{array}{ccc} -1 & +1 & -1 \\ -1 & +1 & -1 \\ -1 & +1 & -1 \end{array}$$

The resulting correlations are then given by

correlation of test input against stored '1' = 3
correlation of test input against stored 'I' = −1

Thus the difference, which gives the discrimination, can be regarded as 133%. In a similar way, if a test input of 'I' is applied, the discrimination against '1' is 57%.

This latter method can be referred to as the external inhibition method, since the negative inhibitory signals are derived from the external input signal rather than from the stored signals in the memory. Not only does this external inhibition method give the best discrimination, but it also requires a minimum size of memory storage and it produces the correlation factor in only a single multiplication cycle on the computer.

A program, known as Mk 17, has been written to compare the three methods discussed with a method using no negative inhibitory signals at all. Six paper tapes were made, by means of a scanning retina, from different sets of characters each containing 37 different characters.

Results with the Mk 17 program

Very extensive tests have been carried out with the computer and artificial retina, using the Mk 17 program. Only an outline of the results will be given here.

The method of external inhibition was found to give the best results in practice, though there was not a great deal to choose between the four methods. The method of external inhibition is also to be preferred because it gives economy of storage.

In tests with these systems, it was found that, in general, the smaller patterns were favoured over the larger ones. For example, a '4' could easily be misread as a '1', especially if it was mis-centred. In such a case the system is extracting the central stroke of the '4' while the rest of the pattern is ignored. Fisher[17] suggests that the reason for this is that too much weight is being given to the negative, inhibitory inputs. In tests with reduced weights of inhibition, making use of various reduction factors, it was found that the best performance was obtained if the correlation sum derived from the negative inputs was divided by a factor of 3. When this was done, the 'four' character was correctly recognised, for example. Fisher points out the very interesting fact that coincidentally this appears to be just the factor used in the human nervous system, judging from the average relative proportions of excitatory nerves to inhibitory nerves[8].

Incoherent Fourier transformation

One approach to visual recognition of characters, which has been investigated at the University of Aston by the physicists Liefer, Rogers and Stephens[9], involves the use of an incoherent form of Fourier transformation. A distinction was made between character recognition, where the members of the class of characters to be recognised are all known *a priori*, and pattern recognition, where the patterns are not known *a priori*, the former being regarded as the easier problem and therefore the one to be tackled.

Methods involving the use of Fourier transformations have the advantage over the use of direct or auto-correlated images that high-frequency noise, such as that introduced by, for example, serifs or minor imperfections, can easily be filtered out, though it must be remembered that in some cases—for example, in the case of capital 'Q'—it would be quite wrong to do this, since it would lead to incorrect recognition.

However, with Fourier transformation it is necessary to introduce some discrimination of phase if it is to be possible to distinguish between characters which are simple mutual inversions (such as '6' and '9').

In practice it has been found possible to discriminate between the ten printed numerals, in a given fixed type-face, by use of three Fourier coefficients measured at seven levels (three bits), only one of these being phase-sampled. Thus ten bits are used, compared with a minimum of 3.3 bits, so that system redundancy can be regarded as equal to 3.

In the experimental equipment the character to be recognised was illuminated by a lamp energised by d.c.; then the image was projected on to a diffusing screen in the focal plane of a telephoto aircraft camera, which was used as a collimator to give parallel rays. A circular scan was then produced by a rotating mirror, so that the overall performance was not affected by the orientation of a following composite shadow transparency grid, of light and dark lines, which was in front of the photo-cell detectors. A second grid was later fitted in front of the photo-cells, to enable larger cells to be used to detect the resulting moiré pattern. In order to detect phase, a reference signal from a spot of light was added.

The resulting experimental equipment gave a 12 bit form of output display on an array of lamps, only 10 of the possible 8192 display patterns being needed. However, since two of the display lamps are never illuminated, they are redundant, and this leaves only 2048 possible display patterns.

Indeed, it is not difficult to eliminate some of the further redundancies in this system if required, and by doing so the present writer has found that it is possible to reduce the necessary display with this system to only five binary places or lamps, i.e. one more than the theoretical minimum, as in *Figure 11.1*. If this is done, redundancy can be re-added as desired in order to improve the reliability of recognition from an engineering viewpoint.

A scanning artificial retina

It is possible to simulate the principle of the ASTRA associating machine by making use of a digital computer. This has been demon-

Character Recognition 207

strated by using a PDP9 computer programmed by Fisher[17] when working at Aston. In this form the ASTRA principle was applied to learned visual character recognition, an artificial retina array of photo-cells being used as the sensory organ.

For use with the computer the input information must be in a serial form, and so it must be converted to this form from the parallel form in which it exists at the artificial retina. It was decided to use a uniselector as a scanning device, since this leads to a much simpler engineering solution than the use of static electronic scanning means. The speed of scanning is unimportant in an investigation of this kind, and uniselectors are readily available.

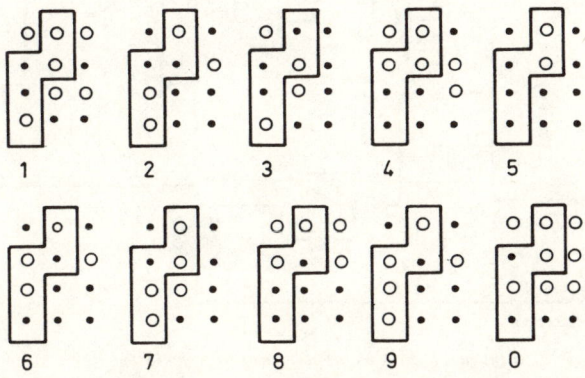

Figure 11.1   Full 12 bit display, with possible reduced 5-bit display enclosed

The selection of a uniselector as a scanning means makes it easy to scan 50 points on the artificial retina, and this fact makes desirable the use of an array of photo-cells of size 7 cells × 7 cells. This uses 49 of the 50 contacts, leaving one spare contact as a resting or end-indicating position.

The circuit arrangement adopted for the scanning device is shown in *Figure 11.2*, In order to provide a resetting facility, a push-button is included in series with the self-stepping contacts, the contact bank and the uniselector coil. When this reset button is pressed, the uniselector rotates in its automatic mode until it reaches the homing position provided by the spare contact, which is not connected.

The uniselector can also be rotated under control of the computer. This is accomplished by the contacts in the computer relay bank, which complete the circuit of the auxiliary relay in the uniselector unit and then release them. Each time that this happens, the contact in series with the coil of the uniselector is closed and then opened, so that the uniselector moves by one step. This intermediate relay is

Character Recognition

required only because the computer switching rating is insufficient for the uniselector coil.

Large capacitors are connected across all contacts and a miniature neon tube is connected across the coil of the relay in order to limit the voltage spike which is fed back to the computer. It was found best to adopt a supply voltage of 35 V in order to reduce the amount of contact bounce in the automatic mode, which was used for testing.

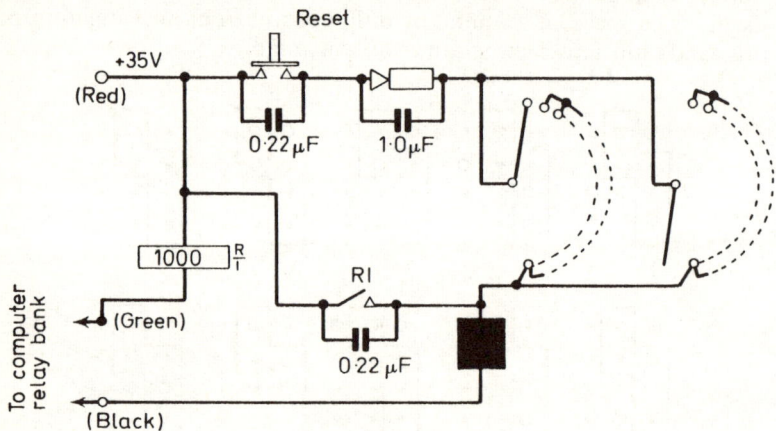

Figure 11.2   Circuit diagram of uniselector drive

The 49 low-cost photo-cells used in the retina each have a resistance varying from 1000 Ω in strong light to 25 000 Ω in darkness. They are mounted in a square matrix of 4.5 cm × 4.5 cm. The sides of the photo-cells are painted black to prevent interference from indirect light, since this has been found desirable in the other applications of these cells described earlier.

The photo-cells are scanned in rows, being each connected in turn to a common 2000 Ω load resistor. The low value of 2000 Ω was chosen in order to ensure the maximum current through the contacts of the uniselector while at the same time providing a large enough voltage signal to operate the following trigger circuit. This trigger circuit of *Figure 11.3* converts the information from the scanner into a two-state form suitable for use by the computer.

Characters drawn by a felt pen on normal duplicating paper, with a line about as thick as the width of the photo-cell, were found to give a satisfactory response, the change in cell voltage being 4–5 V. The trigger circuit was arranged to have a backlash of about 0.5 V, and it gives an output voltage change of about 1.2 V.

With a view to possible practical applications of such a retina—for example, in the reading of postal codes—normal typed characters were enlarged photographically to a height of 4.5 cm. These gave satisfactory results, and it is thus probable that normal optical magnification of typed characters would be satisfactory for use with

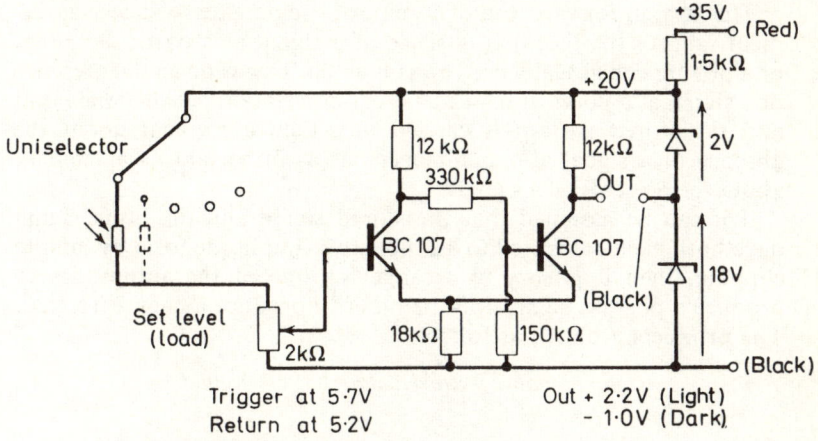

Figure 11.3   Circuit diagram of the trigger output

these low-cost photo-cells. However, all of the experimental work at Aston using this scanning retina has so far used hand-drawn characters written with a felt pen.

This scanning retina has been used at Aston in the simulation of the Astra machine, when applied to optical character recognition, on a PDP 9 computer[12].

Simple pattern-matching gates

It has often been suggested that information which is available in a serial or 'scanned' form can be supplied to a shift register, the outputs from each stage of the register being applied to an AND gate. Then a large output is obtained from the output of the AND gate when there is an agreement between the input scanned information and the master pattern which has been, in effect, set up in the AND gate.

Such pulse pattern detectors were in fact used in some forms of pulse code modulation system[10]. A later development was to replace the AND gate with a majority logic gate. By doing this, it is possible to detect scanned patterns of pulses despite a few errors. Such methods have been proposed for use in visual pattern recognition.

However, there is not really any need for the method of scanning followed by staticisation by a shift register, provided that a number of photo-electric cells in matrix form can be used to detect the visual image rather than a TV camera tube. The photo-electric cells give their output in parallel form, which is suitable for direct presentation to one or more majority logic gates.

The straightforward use of a majority logic gate to detect agreement ignores the fact that absence of a signal at a particular point of a picture can actually convey as much information as the presence of a signal at a point. If instead a cross-correlation between the input and the stored pattern is carried out, then agreement about the absence of a signal at a point becomes as important as agreement about the presence of a signal.

If it can be assumed that the stored signal and the input signal have both been converted to a simple two-amplitude form by infinite clipping, then it is easy to arrange for one of the amplitudes to produce a positive signal and the other to produce a negative signal. The products are then as follows:

Input signal	Stored signal	Product
+	+	+
+	−	−
−	+	−
−	−	+

Thus a positive correlation indicates agreement, whether it is positive or negative. In physiological terms, inhibition is being made use of, and only when there is a disagreement between the stored and the input pattern is there a negative correlation.

REFERENCES

1. Anon., 'Data Entry goes from Strength to Strength', *IEE News*, May 3, 8 (1971).
2. Anon., *Character Recognition*, British Computer Society (1967).
3. Young, John F., *Cybernetics*, Iliffe (1969).
4. Nievergelt, J., 'Review of Perceptrons (Minsky and Papert)', *Trans. IEEE*, **C18**, June, 572 (1969).
5. Hay, J. C., et al., 'The Mark 1 Perceptron-Design and Performance', *IRE Int. Conv. Rec.*, **8**, pt 2, 78 (1960).
6. Taylor, W. K., 'Pattern Recognition by Means of Automatic Analogue Apparatus', *Proc. IEE*, **106B**, 198 (1959).
7. Taylor, W. K., 'Machines That Learn', *Sci. J.*, **4**, October, 102 (1968).
8. George, F. H., *Cybernetics and Biology*, Oliver and Boyd (1965).
9. Liefer, I., Rogers, G. L. and Stephens, N. W. F., 'Incoherent Fourier Transformation: A New Approach to Character Recognition', *Opt. Acta*, **16**, 535 (1969).
10. Tyler, J., 'Basics of Pulse Code Modulation Telemetry', *Orbit*, **6**, April, 23 (1971).

11. Young, John F., 'Cybernetics, Past, Present, Future', *Br. Comm. Electron.*, **12**, May, 302 (1965).
12. Young, John F., *Cybernetic Engineering*, Butterworths (1972).
13. Anon., 'Character Recognition', *Data Processing*, **12**, May/June, 250 (1970).
14. Holmes, W. S., 'Automatic Photointerpretation and Target Location', *Proc. IEEE*, **54**, December, 1679 (1966).
15. Schuh, J. F., 'What a Robot Can and Cannot Do', in: Rose, J. (ed.), *Survey of Cybernetics*, 29, Iliffe (1969).
16. Wee, W. G., 'A Survey of Pattern Recognition', *Proc. 7th IEEE Symp. Adaptive Processes, U.C.L.A.*, *1968*, 2el.
17. Fisher, K. P. 'A Digital Simulation of Astra Mk 3 and Some Experiments in Character Recognition', University of Aston (1970).
18. Carsbury, W. H., 'Electronic Reading Machines', *Telecommunications.* **7**, September, 14D (1972).

12

Robot vision

Minimal character recognition

Very complex equipment is often seen which has been designed to recognise typed or written characters and to convert them into a coded form suitable for use with a computer. Such work is rightly regarded as a part of cybernetic engineering. While such equipment is often extremely complicated, it is of interest to attempt to reduce it to its essentials. Much of the complication lies in the basic mechanical handling systems necessary to feed through the documents which are to be read by the machine. However, the actual reading part of the equipment is often very complex also.

In the Cybernetics Laboratory at Aston, D.G. Hopkins has endeavoured to find methods of reducing the problem of character recognition to essentials[35]. It is usual for the basic type-face to be used with any character recognition system to be rigidly specified in advance. If this is done, then what is the minimal amount of equipment which can be used to recognise, and to differentiate between, a given number of different characters? As with most original work, the answer seems so obvious once it is realised that it might be of interest to follow briefly the process of descent from complexity to simplicity.

Hopkins started by assuming that he would use an array of 25 photo-cells, while recognising that the human eye uses many more sensors than this. Next he devised the shapes of his ten basic numeral characters in such a way that they were composed of straight lines and curves were almost eliminated. In this way it was ensured that the photo-cells used are adequately covered, and ambiguity is reduced. The resulting characters are rather square in shape, but this is not excessive.

The photo-cells are now labelled in rows as row 1, row 2, ..., row 5, and in columns as column A, column B, ..., column E. Any particular photo-cell can then be labelled— for example, as C3. The information about the illumination of each of the photo-cells caused by each of the characters to be recognised was then condensed on to charts. The charts were then examined carefully for redundancies— for example, for cells which were never illuminated and for recurring patterns.

Some cells are found which are covered by all but one of the characters, or else uncovered by all but one. Nine cells such as this could be used to indicate all ten characters (the tenth being made redundant by elimination). However, even if nine such cells could be found, the number is still considerably above the absolute minimum of four cells. The minimum number of cells is four for detection of ten different characters, since $2^4 = 16$. With three cells the maximum number of different characters which could be separated is $8 = 2^3$. Now if cells are found which are covered, or uncovered, by two of the characters, then a minimum number of six cells would be required. If the number of characters allowed to cover or uncover any particular cell is three, then only five different cells are theoretically required.

Following such a line of argument, it is concluded that interest must be confined to those systems in which either four or five of the ten numeral characters can cover or uncover any particular photo-cell.

Working from the reduced chart discussed above, we find by examination that there are three photo-cells in the complete matrix of 25 which meet the specification: namely cells numbered A4, B3 and E2. However, these three cells leave confusion between the numbers 2 and 8 and also between numbers 3 and 9. This is to be expected, since only three photo-cells are used. With the addition of one more photo-cell it should be possible to separate 2 from 8 and to separate 3 from 9. This is so, and examination of the matrix shows that the required additional cell is number A2. Thus with the four cells, A2, A4, B3, E2, it is possible to differentiate between all ten different characters.

Now it is important to note that the configuration obtained is not necessarily unique, specifically because a particular character set was chosen before the required layout of photo-cells was devised. Nevertheless, the demonstration by Hopkins of the feasibility of this approach is most important, since it demonstrates the practicability of a formal design approach to character design and to character recognition. The practical equipment built at Aston by Hopkins is very simple and very successful.

Extension of the minimal method

The original minimal method of numeral recognition described gives a four-place binary-coded output. It would be very desirable if the output could in fact be in a true binary form. Once Hopkins had demonstrated the practicability of this approach, further work by the writer led to such a system which gives a true binary output.

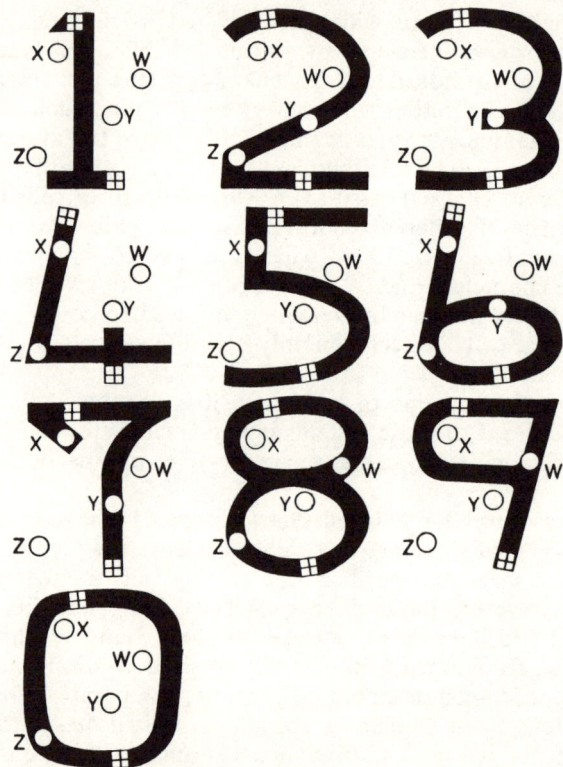

*Figure 12.1   Character set giving a true binary output directly from four photo-cells
+ Marks location points common to all characters
o Marks position of photo-cell*

The form of the characters used in this system and the required layout of the photo-cells are shown in *Figure 12.1*. The binary output from this system is obtained by simple inversion of the digital voltage from the photo-cell Z, so that the output can be written as W X Y $\overline{Z}$.

In any practical system for use in robot machinery or in character recognition equipment, the full minimisation would not be used. The reason for this is that the introduction of additional redundancy of information can help to improve the reliability of the system. The particular form which the redundancy should take will, of course, depend on the nature of the over-all system.

As an example of the introduction of redundancy, the points shown as crosses (+) in *Figure 12.1* are in the same position on all ten of the characters, and so they can be used for location of the characters in the reading equipment.

By extension of his system, making use of photo-cells having the effective shape of long lines (and composed in practice from lines of circular cells), Hopkins has obtained about 90% correct recognition of handwritten numbers of indeterminate size, still using only four cells.

There has been great interest in the possibility of the recognition of handwriting for use with computers, and some of this work has reached a commercial stage. For example, with one system[31] it is possible to write-in information either on a 'Rand' tablet, consisting of a wire grid, with a special stylus, or by use of a special spark pen, in which the sound of the sparks, occurring at a rate of 200/s, is picked up by X–Y microphones at the edges of the board. With this system it is claimed that it is possible to recognise 100 different characters written by any particular person and there is feedback from a cathode ray tube display to the writer, so that he can erase faulty information either by pressing an erase button or by simply scribbing over the faulty character.

Edge detection at the retina

Physiological investigations have indicated some of the features of animal methods of visual detection of images. Lettvin *et al*[1, 17] found that there appeared to be various specialised nerve fibres coming from the eye of the frog, including some which respond only to sharply defined boundaries between objects. Only the images of changing light patterns and of moving curved edges falling on the retina cause signals to be sent to the brain. All other forms of neural information at the retina appear to be ignored and not to cause signals to be sent to the brain. Hubel and Wiesel[2] have discovered that certain cells in the eye of a cat are responsive only to movement of the image on the retina. This fact implies that a direct neural response is possible to the velocity with which an image moves across the retina.

It is seen that to some extent the visual detection of the nature of images takes place directly at the retina and not in the brain. There is thus a great deal of engineering work waiting to be done in simulation of such facilities of the eye and in the adoption of such facilities for engineering purposes. In general, such work has in the past been limited to the use of digital computer processing.

In some cases very complex theoretical guesses have been made about the possible organisation of the contrast enhancement feature of the retina[13, 14].

Contrast enhancement in a scanning system[3, 4, 32]

In order to incorporate bidirectional contrast enhancement in an eye arrangement which incorporates a scanning system, such as, for example, an eye using a TV camera tube[34], it is necessary to use two separate systems.

Consider a television type of scan. Contrast enhancement can take place horizontally, along each line. It can also take place vertically, from each line to each adjacent line. The methods used for the two different forms of enhancement are basically similar. The time-varying signal obtained from the scan is delayed by various amounts of time and the delayed and undelayed versions are added and subtracted in certain proportions.

As an example, consider the vertical enhancement circuit shown in block form in *Figure 12.2*. The process here produces an output

$$(L_{n-1} - (L_n + L_{n-2})) k + L_{n-1}$$

where L_n is the output from scanning line n, etc., and k is a constant. The resulting output can be written as

$$-kL_n + (1-k) L_{n-1} - kL_{n-2}$$

This form of contrast enhancement can be used vertically, when the delay required is equal to the time-length of one line. It can also be used horizontally, when the delays are only required to be of the order of one element length. In the former case accurate delay lines are necessary; in the latter case simple passive circuits can be used, though with the addition of phase inversion amplifiers. The method is similar to that described elsewhere[36].

Although TV pick-up tubes seem to be ideal for use in the robot eye system, there are various problems. They must be carefully protected from damage such as that caused by excessive light. In general, the life of a camera tube is very limited, and this makes them undesirable for continuous use in the robot eye system. Some tubes suffer from a 'burning-in' of the picture on to the sensitive surface if they are exposed to a fixed bright scene for a long time.

Although camera tubes are much cheaper now than was the case a few years ago, there is still room for a cheaper system which can give an increased working life. Sometimes a robot must work in an environment where it must provide its own illumination, and in such instances the flying-spot system is worth considering.

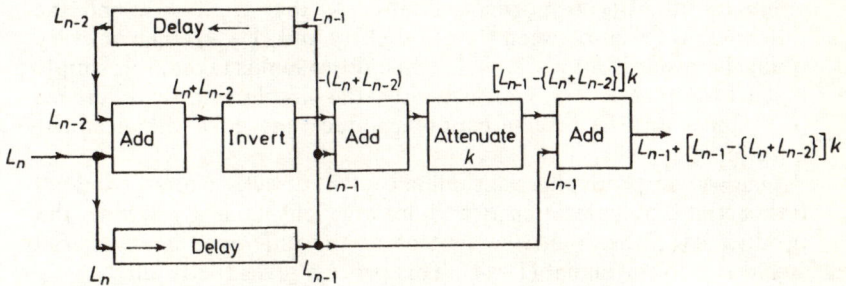

Figure 12.2 Enhancement circuit

The writer produced such a system some years ago for use in hazardous industrial environments where the restricted life of normal camera tubes was an embarrassment. A projection type of TV tube provided a visual raster which was projected on to the scene to be viewed. The light reflected from the scene was then picked up by a photo-multiplier tube. The initial results with this system were disappointing, but the substitution of a projection tube having a special phosphor gave greatly improved results. Unfortunately the high voltages required and the poor availability of the tubes make this system unsuitable for use in present-day systems.

Integrated retinal systems containing many photocells are now becoming available.

Outline enhancement by superposition

The outlines of a visual image can be enhanced by the exact superposition of two opposing visual presentations. For example, the exact superposition of positive and negative transparencies of a picture can give an outline of the picture. Owing to interference and leakage effects at the edges of the images, light can only get through the composite arrangement at regions of sharp change. It has been shown that a certain amount of texture information as well as of outline information is preserved in such a process. In such a process the smaller the spacing between the two transparencies, the finer the detail which can be reproduced.

Consider a purely black and white image. At the edges of the image, the junctions between the black and the white sections, there is inevitably a more or less gradual change in the transmission of light. Now if the simple sum is taken of the intensity A at any given point and its inverse $-A$, then the sum is always equal to zero and there can be no net transmitted image at all. Similarly, if the maximum value of intensity is normalised at 1, so that $A_{\max} = 1$, and the difference is taken between the A intensity and $(1-A)$, then at every point the result is simply $2A - 1$; while if the sum is taken, it is simply equal to $A + 1 - A = 1$. Consequently, the simple sum or difference of a signal and its inverse cannot produce any edge enhancement effect at all.

One way of producing edge enhancement directly from the images rather than by interference and leakage effects is by taking the product of, rather than the sum of or the difference between, the signal A and the signal $(1-A)$. This process gives the result

$$A(1-A) = A - A^2$$

A resultant signal of this form has the effect of producing a maximum value of output at the point where the intensity of the image has one-half of its maximum value.

The normalisation process can be carried out by division by the maximum value, leading to the product signal

$$\frac{A}{A_{\max}} - \frac{A^2}{A^2_{\max}} = \frac{A}{A_{\max}}\left\{1 - \frac{A}{A_{\max}}\right\}$$

Here $A_{\max}$ must be defined as the value at the nearest point of zero slope and maximum negative rate of change.

It is not necessary to take the base of the calculation as zero if the nearest point of zero slope and maximum positive rate of change is also determined, since an indication of local slope can then be taken as

$$\left(\frac{A - A_{\min}}{A_{\max}}\right) - \left(\frac{A - A_{\min}}{A_{\max}}\right)^2$$

where $A_{\min}$ is the value at the nearest point of zero slope and maxi- positive rate of change.

This is an entirely local feature. It will be noted that, unlike the process of scanning differentiation, the process described above is independent of the rate of change and depends only on the actual values. It has the disadvantage that it gives a large output for a slow change, since it depends on actual values rather than on rates of change.

It is of interest to note that the addition of two apparently completely random arrays of dots can form a very real image. An excellent illustration of this is given by Schroeder[12]. Work such as this illustrates very forcibly the difficulty involved in attempting to make some sense out of photomicrographs of arrays of neurons found in nature.

A counting retina

In order to count a large number of objects using his eyes, a human being has to count them individually or divide them into smaller groups. However, the human eye is capable of an instant counting perception of the retinal images produced by a small number of objects[39, 40]. Such a capability in an artificial retina would have a number of applications, a notable example being the rapid determination of the number of objects, say blood cells, on a microscope slide.

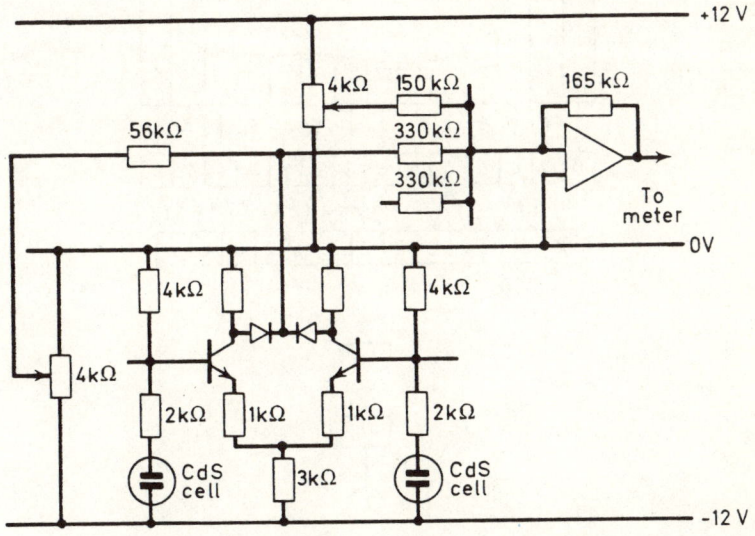

Figure 12.3 Apparatus for direct counting retina

The basic possibility of constructing such a direct counting retina has been demonstrated by P. S. Williams at Aston[37]. The experimental apparatus is shown in *Figure 12.3*. A single line of cadmium sulphide photo-electric cells represents a line of retinal cells. The

220 Robot Vision

signals from these cells are taken to the processing amplifiers and then to the output. The number of objects appearing before the line of cells is determined by the simple circuits and the result is displayed on a meter, the scale of which is marked with the number of objects.

The principle of operation adopted for this demonstration equipment is very simple indeed, although it sounds like a joke at first hearing. In order to determine the number of objects, regardless of their size and position, the number of edges of objects appearing before the retina is counted, and then the circuitry automatically divides the number of edges by 2. Thus the principle of edge detection is used in this form of artificial retina.

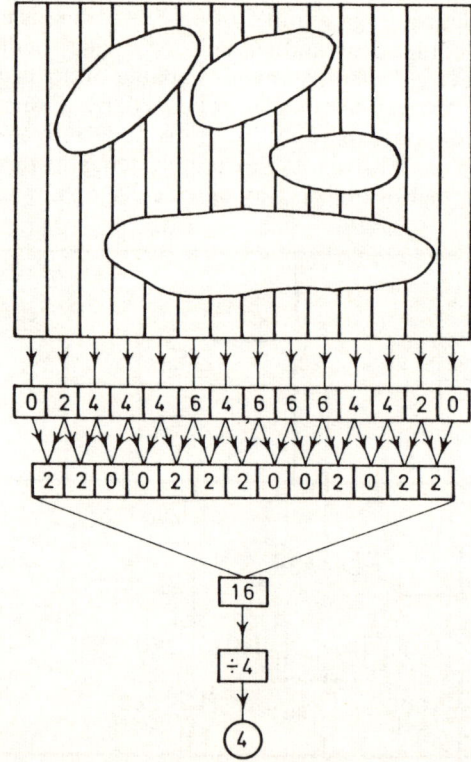

Figure 12.4   Using the Aston edge detector

The principle can be extended for use in two dimensions, rather than only one, simply by using an Aston edge detector only to determine the number of edges along each of a raster of parallel lines in the visual scene. Differences are then taken as in *Figure 12.4* between the number of edges appearing in adjacent strips. The differences are then summed and divided by 4, to give the total number of convex objects being viewed, regardless of their individual sizes. Such a counting device has direct application for counting objects in industrial stores or in shops, in addition to the medical applications mentioned on p. 219.

It should perhaps be mentioned that there are limitations in the use of the simple counting scheme described. There must be a complete line of illuminated photo-cells at each end, though these can be simulated. Adjacent objects must either be separated by a complete row or overlap on the same row. A form of optical illusion miscount occurs if two object boundaries occur on two separate but contiguous rows. The objects must not be concave, and a miscount can occur unless convex objects are counted. Despite the deficiencies, the method illustrates the results obtainable with an arrangement which is very simple in principle. The method is, of course, directly applicable with a scanning device such as a camera tube instead of the discrete photo-cells used.

An edge detecting retina

From various investigations of animal systems, as mentioned above, it is known that the detection of edges plays an important part in recognition. As an example of this, consider the 'learner' plate consisting of a red L on a white background which is not uncommon on automobiles in Britain. All of the red area inside the L and all of the white area outside conveys little information. It is the configuration of the edges forming the junction between the red area and the white background which indicates that the figure in an L and not, for example, a T.

A modification of the counting form of retina described above could be used to detect edges and to give an indication of their positions. However, if an extensive retina is required, using large numbers of photo-electric cells, then the necessary number of differential amplifiers would be excessive and expensive.

While working on the counting retina in the Aston Cybernetics Laboratory, C. E. Free[38] had the ingenious idea of adopting an alternating supply voltage, rather than d.c. as was used in the earlier work. This has made it possible to construct edge detecting retinas requiring the use only of photo-cells, of capacitors and of resistors, no transistors or other active devices being necessary at all.

The basic arrangement is shown in *Figure 12.5*, which represents a single line of retinal cells.

222 Robot Vision

Consider photo-cells C and D. If neither of these photo-cells is illuminated, then there can be no output voltage at point X. On the other hand, if both photo-cells C and D are equally illuminated, then a voltage appears across the resistor R_3 on each of the half-cycles. There is consequently a symmetrical alternating voltage across R_3.

Now if the product $R_c C_c$ is large compared with the cycle time of the alternating supply, then the symmetrical alternating voltage across R_3 will be smoothed out almost completely and again there will be little or no output voltage at point X.

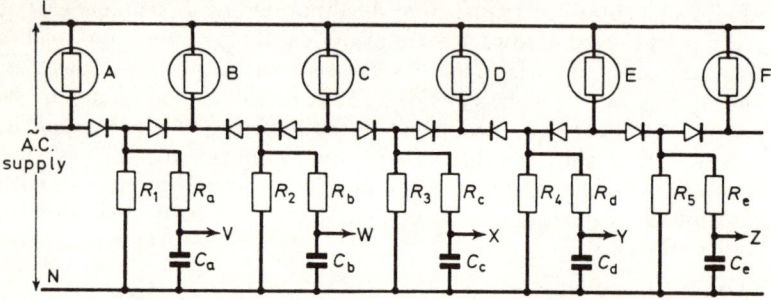

Figure 12.5 Basic circuit for an edge detecting retina

Now consider photo-cell C to be illuminated but photo-cell D not to be illuminated. This will be the case if, for example, there is a dark edge which obscures photo-cell D but does not obscure photo-cell C, i.e. there is an edge which falls in between photo-cells C and D. In this case, on the supply voltage half-cycle when the supply line L is positive to the supply line N, photo-cell C will conduct and a positive voltage will appear at output point X. On the other hand, on the supply voltage half-cycle when the supply line L is negative to the supply line N, there is much less conduction through photo-cell D, since it is dark. Therefore little output voltage appears in the negative direction. In consequence, the output capacitor C_c charges much more in the positive direction than in the negative direction, and a positive smoothed output appears at point X.

In a similar way, if cell C is not illuminated but cell D is illuminated, then a negative output voltage appears at point X. The operation of this arrangement can therefore be expressed by the following table:

cell C	cell D	point X
dark	dark	zero volts
dark	light	negative volts
light	dark	positive volts
light	light	zero volts

This arrangement thus only gives an output voltage at any point when it is actuated by an edge at that particular point.

This form of circuit is easily extended for use as a full two-dimensional edge detecting retina. Neon indicator tubes are used for display purposes, and they light only along a line when, for example, the edge of a card appears in front of the retina of photo-cells.

Free's original arrangement used a sinusoidal alternating supply, but later developments for use with the Astra machines are making use of pulse-type power supplies and they give outputs compatible with the forms of logical circuitry used in these machines.

Future possibilities for retinas[5-11, 15, 16, 33, 41]

There have been various attempts to produce an integrated-circuit form of artificial retina, particularly for use in computer card scanning. One of the problems encountered has been the fact that while it is possible to make an array of photo-electric cells of very small over-all size, it is very difficult to bring out the connections from the cells. In some cases it has been necessary to reduce the retina to a single line of cells and to have in effect a sequential read-out from the cells. Even so, such devices have been very expensive, presumably because of the small demand and the necessity for all of the cells in a line to be successful operating devices, with no failures permitted at all.

A newer form of retina has recently been produced by R.C.A. This comprised a total of 960 photo-sensitive elements in a flat array, but unlike the usual micro-miniature integrated circuit, the newer device is deposited on a sheet of glass measuring 10 cm × 20 cm. Each of the photo-cells is of the lateral-flow type and is connected to mutually perpendicular terminal strips via a thin-film Schottky diode. The stages.
whole device is made by vacuum deposition in a number of

The OPT 5 integrated circuit has a 10 × 10 array of photo-diodes, together with scanning circuits. It is interesting to note that there is a large change in the Fourier coefficients with pattern movement if such an array with a small number of elements is used.

Weimer and others[7] have described the extensive work carried out on the use of self-scanned retinal sensors using integrated circuit techniques. Arrays of many thousands of photo-transistors have been proposed for use in planetary exploration. In these the collector-to-base capacitance is used to integrate the light flux, and it is discharged once per frame. Other devices have been described[42, 43].

Edge detection in a computer scanning system

A standard Vidicon TV camera is used in the Stanford hand-eye system, which uses a robot arm to build a stack of visually detected cubes. In the early work[19], black cubes were used on a white table, and the system operated satisfactorily only if the contrast level was high. Sixteen levels of quantisation were used, and even with human tuning for best results it is not always possible to resolve all edges simultaneously in a complex scene. Consequently, both the camera and the lighting level had to be tuned manually by a human operator in order to obtain the best possible resolution of the particular object under consideration at the moment. Naturally, this is not a satisfactory approach, since the results obtained might well be influenced by the human intervention. The program of the computer would automatically reject any data which did not indicate a reasonable number of corners or a satisfactorily closed outline of an object[21]. Later work has improved on this[20, 18].

The target voltage of the Vidicon camera tube is automatically adjusted to get best results while at the same time ensuring that the voltage is not sufficient to damage the tube. The picture is automatically focused by moving the tube with reference to one of the lenses from a multiple turret. It was suggested that colour filters could be used in a turret, being tried at random to obtain the best contrast. The Hueckel form of regional operator[22], which can detect edges even if they are weak and there is substantial noise, is used, and the computer then traces round the edge of the image of the object while recording the lines and end-points.

The program used builds up the information about the blocks being examined as it is acquired. For example, if a previously encountered point is reached during the tracing procedure, then the stored data are examined, to check, for example, whether a closed outline has now been traced. All of the edges of any particular object need not, therefore, be traced consecutively.

The earlier edge follower often did not complete a contour if one small section was noisy or difficult. The newer program can cause a jump to occur over any particularly bad area, or it can attempt to close the contour in the opposite direction. Finally, various 'tidying-up' procedures are followed: for example, to make all edge points meet at corners.

Other methods which have been tried include the use of digital spatial filtering techniques for image enhancement[23], the use of high-level context to infer missing detail by syntax-directed analyses[24, 25] and the use of regions rather than edges[26], though it has been stated that in general, in these methods the image to be analysed was fixed and the computer could not alter the action of the camera or scanner.

In the Stanford equipment a standard Vidicon camera was used with motorised computer control of the pan–tilt head, the lens turret, colour filter, focusing and target voltage with manual setting of the iris. Three colour filters and a neutral density filter are fitted, on a filter wheel, which allows selection in 0.2 s. There are 64 possible target voltages between 0 and 50 V, though the voltage is not permitted to cause an excessive average signal current.

Sixty times each second, the Vidicon scans an array of 333×256 samples of intensity, each of which is encoded as a 4 bit number so that the 24 million bits/s capacity of the high-speed data channel of the computer is not exceeded. However, the range of signal voltage represented by the 4 bit, 16 level number can be varied from the full 1 V working range of the video amplifier down to a 'sub-window' range of only $\frac{1}{8}$ V (thus giving an effective range of 128 levels).

One technique which has been suggested for simplifying the computer or robot recognition of three-dimensional shapes is that known as 'grid coding'[27]. In this the scene is illuminated but by a uniform light source but by a projected grid of lines, while the viewing camera is offset somewhat from the light source. It has been suggested that this technique might give improved results compared with the earlier methods, all of which use the work of Roberts[28]. A similar method to that of grid coding has been used in Japan, where a single moving slit has been used to illuminate the object to be viewed[29]. The same workers have used successive illumination of a scene from different directions in order to extract the information for production of a line drawing of the scene. Stereoscopic viewing with two cameras as used at M.I.T. was not used by the Japanese, because of the extensive processing required in order to determine the shapes being viewed. The Japanese workers have also used sequential illumination of a scene from several directions in order to obtain the information for construction of a line drawing by computer[30].

REFERENCES

1. Lettvin, J. Y., *et al.*, 'What the Frog's Eye Tells the Frog's Brain', *Proc. IRE*, **47**, 1940 (1959).
2. Hubel, D. H. and Wiesel, T. N., 'Receptive Fields, Binocular Interaction and Functional Architecture in the Cat's Visual Cortex', *J. Physiol.*, **160**, 106 (1962).
3. Dennison, R. C., 'Aperture Compensation for Television Cameras', *RCA Rev.*, **14**, December, 569 (1953).
4. Sierra, H. M., 'Bit Shift and Crowding in Digital Magnetic Recording', *Electro-Technology, N.Y.*, **78**, September, 56 (1966).
5. Evans, F. C., 'Use of a Solid-State Scanner for Pattern Recognition by Fourier Analysis', *Electron. Eng.*, **41**, November, 43 (1969).

6. Weimer, P. K., et al., 'Multi-Element Self-Scanned Mosaic Sensors', *IEEE Spectrum*, **6**, March, 52 (1969).
7. Weimer, P. K., et al., 'A Self-Scanned Solid-State Image Sensor', *Proc. IEEE*, **55**, September, 1591 (1967).
8. Salomon, P. M., 'Application of Slow Scan TV Systems to Planetary Exploration', *J. Soc. Motion Picture Television Engrs*, **79**, 607 (1970).
9. Assour, J. M. and Lohman, R. D., 'A Photodiode Array for Holographic Optical Memories', *RCA Rev.*, **30**, December, 557 (1969).
10. Lohnson, R. E., 'Vidicon Performance in Extreme Thermal Environment', *RCA Rev.*, **27**, 360 (1966).
11. Plummer, J. D. and Meindl, J. D., 'A Reading Aid for the Blind Using MOS Electrodes', *Proc. 23rd Conf. Eng. Med. Biol.*, Vol. 12, 168 (1970).
12. Schroeder, M. R., 'Images from Computers', *IEEE Spectrum*, **6**, March, 66 (1969).
13. Deutsch, S., 'Conjectures on Mammalian Neuron Networks for Visual Pattern Recognition', *Trans. IEEE*, **SSC2**, December, 81 (1966).
14. Sutro, L. L. and Kilmer, W. L., 'Assembly of Robots to Command and Control a Computer', *A.F.I.P.S. Conf. Proc.*, **34**, Spring, 113 (1969).
15. Anders, R. A., et al., '100 x 128 Element Solid State Imaging System', *Wescon Conv. Rec.*, pt 1, paper 13/1 (1967); Weckler, G. P. and Dyck, R. H., 'Integrated Arrays of Silicon Photodetectors for Image Sensing', *Wescon Conv. Rec.*, pt 1, paper 13/2 (1967).
16. Hlwatashi, K., 'The Quest for a Seeing Machine', *New Scient.*, **32**, November 3, 232 (1966).
17. Herscher, M. B. and Kelley, T. P., 'Functional Electronic Model of the Frog Retina', *Trans. IEEE*, **MIL7**, April/July, 98 (1963).
18. Pingle, K. K. and Tenenbaum, J. M., 'An Accomodating Edge Follower', *Proc. 2nd Int. Joint Conf. Artificial Intelligence, September, 1971* (Br. Computer Soc.), 1.
19. Pingle, K. K., Singer, J. A. and Wichman, W. W., 'Computer Control of a Mechanical Arm through Visual Input', *Proc. I.F.I.P.S.*, 1563 (1968).
20. Tenenbaum, J. M., et al., 'A Laboratory for Hand-Eye Research', *Proc. I.F.I.P.S.* (1971).
21. Pingle, K. K., 'Visual Perception by a Computer', in: *Automatic Interpretation and Classification of Images*, 277, Academic Press (1969).
22. Hueckel, M., 'An Operator Which Locates Edges in Digitized Pictures', *J. Ass. Computer Machinery*, **18**, January, 113 (1971).
23. Rosenfeld, A., 'Picture Processing by Computer', *Computing Surveys*, **1**, September, 147 (1969).
24. Miller, W. and Shaw, A., 'Linguistic Methods in Picture Processing—A Survey', *Proc. Fall Joint Computer Conf.*, 279 (1968).
25. Fischler, M., 'Machine Perception and Description of Pictorial Data', *Proc. 1st Int. Joint Conf. Artificial Intelligence, Washington, 1969*.
26. Brice, C. and Fennema, Cl. 'Scene Analysis Using Regions', *Artificial Intell.*, **1**, 1 (1970).
27. Will, P. M. and Pennington, K. S., 'Grid Coding: A Preprocessing Technique for Robot and Machine Vision', *Proc. 2nd Int. Joint Conf. Artificial Intelligence, September, 1971* (Br. Computer Soc.), 66.
28. Roberts, L. G., 'Machine Perception of Three-Dimensional Solids', in: Tippet, J. T., et al. (eds.), *Optical Information Processing*, M.I.T., (1965).
29. Yoshiaki Shirai and Motoi Suwa, 'Recognition of Polyhedrons with a Range Finder', *Proc. 2nd Int. Conf. Artificial Intelligence, September, 1971* (Br. Computer Soc.), 80.
30. Yoshiaki Shirai and Saburo Tsuji, 'Extraction of the Line Drawings of

3-Dimensional Objects by Sequential Illumination from Several Directions', *Proc. 2nd Int. Conf. Artificial Intelligence, September, 1971* (Br. Computer Soc.), 71.
31. Anon., 'Program Recognises Handwritten Math.', *Electronics*, **44**, August 20, 26 (1971).
32. Khol, R., 'Image Enhancement', *Mach. Des.*, **44**, April 20, 181 (1972).
33. Parrish, E. A. and McVey, E. S., 'On the Feature Selection Problem of a Reading Machine for the Blind', *Trans. IEEE*, **IM21**, May, 168 (1972).
34. Wall, W. A. and Stephens, D. L., 'Automatic Closed Circuit Television Electrode Guidance for Welding', *Welding J.*, **48**, September, 713 (1969).
35. Hopkins, D. G., 'Optical Character Recognition', University of Aston (1970).
36. Young, John F., *Cybernetic Engineering*, Butterworths (1972).
37. Williams, P. S., 'Object Counter Using an Artificial Retina', University of Aston (1968).
38. Free, C. E., 'Construction of an Artificial Retina', University of Aston (1968).
39. Miller, G. A., 'Information and Memory', *Sci. Am.*, **195**, August, 42 (1956).
40. Miller, G. A., 'Seven Plus or Minus Two', *Psychol. Rev.*, **63**, March, 81 (1956).
41. Gilder, J. H., 'Three Years after Birth, C.C.D.'s Head for First Commercial Applications', *Electron. Des.* **21**, January 4, 36 (1973).
42. Pike, W. S., *et al.*, 'An Experimental Solid State Television Camera Using a 32 × 44 Element Charge-Transfer Bucket Brigade Sensor, *R.C.A. Rev.*, **33**, September, 483 (1972).
43. Anon., The Smaller Cheaper Television Camera, *Electron*, March 22, 7 (1973).

13

Perception of movement

Eye movement

In the course of most engineering work in which attempts have been made to simulate some functions of the animal retina or eye, it has been usually assumed inherently that the pattern to be viewed must be projected into a fixed position on the retina. Positioning has often been accomplished by movement of the object to be viewed into the correct position in front of the fixed retina[23]. However, a number of potential advantages are to be gained by making any artificial retina movable as is the animal retina. The disadvantage of this course is that the speed of operation of such a device could not greatly exceed the speed of operation of the animal eye, and the fixed retina method is probably best if a high speed of operation is required. However, in any humanoid robot it would seem to be desirable to introduce retinal movement.

Each human eye can be moved by six muscles. From an engineering point of view, a minimum of three is required to control swivelling of a ball-shaped eye. It will probably simplify the controls if four muscles are used in two pairs, horizontal and vertical. A further simplification can be obtained in the engineered robot by using return springs which oppose the two, horizontal and vertical, muscle devices. However, it is then necessary to provide a continuous excitation to hold the retina at a central position.

A lot can be learned about the positioning of an artificial retina from the engineering work which has been carried out on the development of two-motion chart recorders, in which a pen has to be moved in two dimensions over paper in order to draw a graph. Attempts have been made here to improve the speed of response of movement of the pen and also the fidelity of reproduction on the graph of the controlling input electrical voltages. With the background knowledge acquired from such work, it can be stated that it is not going to be an easy matter to compete with the movement performance of the animal and human eye. An additional problem, as with all other movement of the robot, is the audible noise produced by the movement.

It is an interesting point that it has been found in work on chart recorders and similar devices that a small amount of tremor in the movement is invaluable. As is well known, static friction is greatly in excess of kinetic friction. Once there is some motion, the friction is reduced. The writer has made use of this in electromechanical servo systems in which a linear movement was required, by introducing a continuous rotation of the rod which had to be moved, the rotation being at right angles to the direction of linear movement. In the animal eye similar small tremors to those required to remove static friction are always present[22]. Here, however, it can be shown that the tremor is a quite essential part of the process of seeing.

Eye tremors[30]

It is a surprising fact that in most of the work on machines for the recognition of visual patterns which has been carried out all over the world, one very simple feature of animal vision appears to have been completely neglected.

This is the continuous tremor, or slight 'Saccades' which occur continuously[1]. In the primitive microscopic creature *Copilia*, movement of the simple retina from side to side is used apparently to economise on the number of retinal cells required. In effect, the creature has a visual system which works in a very similar way to the scanning of a television picture.

In more advanced creatures and in man, the continuous judder of the eye appears to be just as essential to vision as it is in the primitive *Copilia*. It is possible by various means, all somewhat painful, to prevent the optical image falling on the retina from juddering at all. The method used involves, in effect, fastening the picture to be looked at firmly to the eyeball by means of a stick and some suitable adhesive. If this is done, then the picture being looked at fragments into more elementary shapes, such as straight lines, angles, curves and so on. It appears that a possible function of the scanning motion is the coupling of these elemental parts together to produce a meaningful whole. Evans[17] suggests that fragmentation occurs when the information entering the eye is reduced—for example, by reduction of the illumination.

One useful function of optical judder is the reduction or elimination of optical noise. This has been used in engineering equipment incorporating fibre optics as light guides[2]. A bundle of thin glass fibres can be bunched together and used to guide light from one place to another. A picture projected on to one end of the bundle can be seen at the other end of the bundle provided that the fibre array is maintained in the same order at each end. However, because of the finite size of the cross-section of the individual fibres, the received picture is seen as a number of individual areas rather than as a complete whole. The breaking-up of the picture into finite dots in this way can be thought of as a form of optical noise. It is found that the visible noise can be greatly reduced by vibration of the whole bunch of fibres, so that each fibre, in effect, scans a small area of the picture.

The following is a very simple illustration of the elimination of optical noise effects by the introduction of movement. A simple line drawing, say of a face, is carried out in black ink on a transparent medium. On another identical transparent medium random lines are drawn. This 'optical noise' is placed in front of the drawing of a face. Although the human eye is then able to see all of the lines making up the face, the face is not discernible as a unit, because it is mixed up with the similar lines which make up the random optical noise. However, if either the face drawing or the noise drawing is moved, the face immediately becomes visible. If the movement is stopped and the viewer looks away for a few seconds and then looks back at the composite picture-and-noise, the face is again not visible, even though the viewer knows it to be there. The demonstration of the effect of relative movement of picture and background is most impressive.

A well-known illusion studied by McKay is the apparent, though non-existent, lines which appear at right angles to an array of radial lines. Once again, it has been suggested that this illusion disappears if the relative movement between eye and picture is prevented. It is thus possible that the effect is a form of moiré effect between the image as seen and the image as it was a fraction of a second earlier.

It will be seen from the above discussion that the phenomenon of tremors in the eye is likely to be of importance in cybernetic engineering. The fact that a reduction of illumination has the same effect as the prevention of relative judder movement, namely fragmentation of the picture, makes it seem likely that in the artificial retina some form of judder will be required in order to assist meaningful pattern recognition, particularly in conditions of optical noise.

The judder can be applied mechanically in the robot retina. However, it is quite likely that an electronic method of introducing an effective judder scan, using no moving parts, will be preferred.

It is of interest to note that a form of judder or dither in brightness level has been used to reduce quantisation noise in digitally coded video signals[14-16].

Fibre optics[2]

While it is at the present time possible to make photo-cells of very small size, there is a basic difficulty in making use of these in artificial retina systems. This is the problem of connecting the large number of electrical wires which are required to be taken to the individual cells of an artificial retina. While it is possible that as our knowledge of visual processes develops it will not be necessary to connect individual retinal elements to the central brain process, it is likely that there will always be a need for a great many connections.

Some progress has been made in the use of an integrated-circuit approach to this problem, but the solution here has generally been confined to the use of a single line of cells rather than a complete two-dimensional array.

It is therefore desirable at the present time that arrays of larger photo-cells should be used in an artificial retina. The image can then be magnified optically by the use of lenses so that it is large enough for application to the retinal array. However, both the required physical size of the optical system and the mechanical stability which is necessary with such an approach cause problems when they are applied to mobile robot equipment.

An attractive alternative approach is the use of flexible light-guide fibres to convey light some distance from the relatively small 'retina' to a comparatively large array of photo-cells situated in a more convenient place. With this approach there is the additional possibility of applying transverse vibration as mentioned elsewhere.

Baird suggested the use of glass rods as light fibres as early as 1926, though it is doubtful if there was much practical work until the 1950s. Fibre optics use internal reflection in order to obtain transmission of light without excessive loss. This approach requires the use of optical transmission fibre material having a refractive index greater than 1.4.

The internal reflective surface is usually protected from contamination by covering with a glass or plastic coating having a lower refractive index than the fibre. This coating is required to have a minimum thickness of one-half wavelength of the light being transmitted if excessive loss of light is to be avoided. Sometimes an additional black coating is applied over each fibre to prevent optical

232 Perception of Movement

interaction with other fibres. In general, it is difficult to obtain a reasonable optical performance with fibres of very small cross-sectional area.

Transmission fibres will only accept light over an angular range which has a limited angular deviation from the axis of the fibre. This fact is sometimes expressed by a figure known as the numerical aperture, which is equal to the value of the sine of the maximum angle of deviation from the axis of the fibre:

$$\text{numerical aperture} = I_{\text{ext}} \sin\theta = (I_f^2 - I_c^2)^{\frac{1}{2}}$$

where I_{ext} is the refractive index of external medium

$$(I_{\text{ext}} = 1 \text{ for air});$$

θ is the maximum angle to axis at which light is accepted by fibre; I_f is the refractive index of fibre; and I_c is the refractive index of the covering of the fibre. The maximum angle of the axis is usually about 33°, corresponding to a numerical aperture of 0.54. A wider acceptance angle can be obtained if the core glass has a higher value of refractive index and a greater absorption of light.

In general, any bends in the fibre should have a minimum radius of 10 times the rod diameter in order to avoid excessive light leakage. Consequently, the glass fibres or rods should have a small diameter if sharp bends have to be used. It is possible to produce fibres having a diameter as small as 0.01 m.

There are various sources of light losses in a light guide, and the amount of light transmitted therefore falls with the length of the guide. Typically, while 60% of the light is transmitted in a short light guide, only 10% is transmitted to the end of a fibre of 250 cm length. A rod of fibres can be drawn down to a reduced diameter at one end, so that it can give an effective magnification or reduction of image size.

Fibre optic eye

A form of robot retina using fibre optic light guides is the 'Iitri' eye. This uses two intermingled arrays of light guides, one of which is used to carry light from a light source to the object, the other being used to carry light back from the object to a matrix of optical sensors. The output from the photo-sensors is then fed into the electronic recognition unit. The recognition unit rejects the information from the retina if there is excessive light which effectively blinds the eye, or if there is insufficient light for recognition or if there is a circuit or other malfunction.

The effective viewing range of this device can vary from about 1.5 mm to over 5 cm, the range depending on the resolution desired and the nature of the part being viewed. At distances up to about 2 cm displacements as small as about ± 0.75 mm can be detected, and it is hoped to improve on this resolution by improving the optics.

Eventually it is hoped to produce an octagonal eye containing some 1200 optical elements at a quantity price of less than £400–500.

The eye following a moving object

When the eye has to look at an object which is moving within the field of vision, there are a number of methods which can be adopted by the human. When scanning across a fixed object in front of the eye, it is usual to carry out a rapid saccadic movement, followed by a period of rest during which the scene is viewed, followed by another saccadic jump to the next position. This method is used by the human when reading printed matter.

In some cases when a moving object is being viewed, the whole head moves in order to maintain the image fixed on the retina[32]. If the head is restrainted from moving, the mode of action becomes more complex.

If the object being viewed moves predictably, as does, for example, a pendulum, then there is no difficulty in following the movement by rotation of the eyeball in its socket. In fact, in such a case the human control system actually appears to apply a phase lead, of the order of 10°, to the movement of the eyeball as compared with the movement of the target. In addition, the movement of the eye is slightly greater than is the movement of the object being viewed.

However, if the object oscillates more rapidly than about once per second, then the gain of the following system falls off rapidly, and the phase shift rapidly increases. The corresponding transfer function has been expressed as

$$\frac{1.08}{(1+sT)^2 (1+2KsT+s^2 T^2)}$$

where T is of the order of 0.1 s and the damping factor K is about 0.35.

While this expression matches the gain characteristics of the eye-following system well, it does not match the measured phase characteristics. The mismatch can be corrected by the inclusion of a predictive element having a transfer function $\exp(+sT_1)$, where T_1 is of the order of 0.3 s.

234 Perception of Movement

If the movement of the object being viewed varies in an unpredictable manner, or when the head is moved during viewing, the movement becomes more complex, as might be expected, though there have been attempts to describe the action mathematically.

It is usual to assume that the eye of a robot should be mounted in the head, but this is, of course, not at all necessary. There is some advantage in mounting an eye on a movable arm provided that it can be protected from damage. While this course would provide a very flexible form of visual device, the necessary control systems would be most complex and it is unlikely that the approach will be used in the near future. On the other hand, the addition of some form of eye to the normal robot arm would be most useful, provided that the weight and size are small, even if the resolution has to be very restricted. One robot eye was mounted in the hand[35].

For experimental purposes, a human eye movement has been measured by detection of the voltage induced in a search coil, carried on a contact lens, by a magnetic field[3]. A similar method could be used in the robot eye.

Other points on eye movement

It seems that some part at least of the process of seeing is learned in the normal course of growing up. One very interesting fact from our present viewpoint is that people who have lost their sight at a very early age and had it restored surgically as adults have the greatest difficulty in differentiating between visible figures which they are quite able to describe verbally. For example, if a square and a triangle are shown, although it is realised that they are different figures, it is not possible to name the shapes unless they can be physically traced and the number of corners can be counted by touch. Experiments with animals such as kittens make it pretty clear that there is little learning by purely visual means unless the visual image can be related to touch and to movement of the body, making use of the muscles of the animal. It may be that we are attempting too much too early in attempting to build visual learning machines which are as yet incapable of movement and the association of such movement with the visual input. It may be that the muscular movements used in viewing an object are themselves associated with the visual images obtained. It is of interest to note that the edge of the retina, which appears to play an important part in the alignment of the eye on an object of interest, might be of great importance in the pattern-learning process in addition to its function of warning of approaching objects.

Work on the eye movement and the head movement involved in

Perception of Movement 235

centring the required image on the retina will no doubt be helped by the work which has been carried out on the control of a TV camera in response to human head movements. An early apparatus of this type was built by Philco in 1958. This had a relatively wide viewing angle and a poor resolution, but nevertheless it did demonstrate the practicability of this approach[4].

Work was later carried out on a more advanced system at Argonne[5,6]. In this a monitor tube is kept pointing at the operator's eyes despite movements of his head through $\pm 80°$ in pan (side-to-side) and $+45°$ to $-30°$ from the horizontal in a tilt motion. The monitor screen was at a distance of about 60 cm from the face of the operator, while the camera was mounted about 100 cm from the slave tongs of the remotely controlled manipulator being observed. The maximum speed of rotation of the camera and the monitor was about $30°/s$.

The headpiece worn by the operator was counterweighted, so that the operator felt little restraining force. It was found advantageous to include a dead zone of $7-12°$, to remove the operator annoyance caused by persistent movements of the monitor in following minor movements of the operator's head. These minor movements can also cause a blurring of the image due to the time constant of the Vidicon camera tube.

In later work the camera was arranged also to follow movements of the head of the operator in the three translational directions, side to side, up and down and forwards and backwards, dead zones again being introduced to remove annoying minor movements. The angle of view was about $30°$. It was found with this system that the operator could gain a great deal of depth, or three-dimensional information, simply by making movements of his head.

With this later system the maximum travel was $\pm 90°$ pan and $\pm 45°$ tilt, the maximum speed of movement being about $23°/s$. The side-to-side and forward and backward movements both covered a range of about ± 30 cm, with a maximum speed of about 4 cm/s, while the up-and-down movement corresponded to about 30 cm downward only from normal, the maximum speed of movement again being about 4 cm/s.

The performance of devices such as these can help to give a guide to the performance, both required and obtainable, for robot viewing systems.

Walter has pointed out that a decision to move the eyes must be accompanied by some form of blanking process so that the scene does not jerk with eye movements[7]. The process does not operate if the eyeball is moved passively. It will be desirable to incorporate such a blanking process into the robot visual system.

236 Perception of Movement

It is of interest to note that a certain amount of visual searching can be carried out by the brain with no eye movement at all[8].

Rees[18] seems to have devoted considerable thought to the use of a TV camera as a robot eye, and some experimental results were obtained.

Automatic focusing of the robot eye[13]

It is possible that, in addition to its function of reducing the effective noise level, the saccadic movement helps to provide nervous signals for the control of the focusing muscles. Certainly, this type of movement can be used to provide signals from the robot eye which enable a degree of automatic focusing to be used.

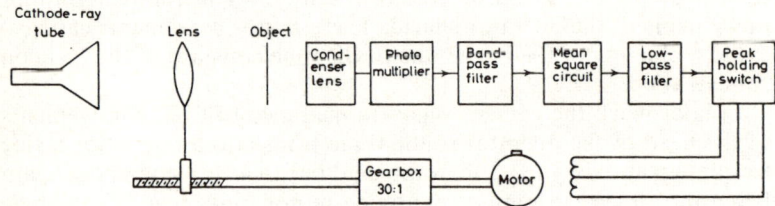

Figure 13.1 Automatic focusing circuit

One system which has been investigated[9] is illustrated in *Figure 13.1*. It is not at present very practicable to use a variable-geometry form of single lens as used in the human eye. Instead a single lens is used for focusing, and this is moved by means of a screw operated by an electric motor servo system. A circular scan of the image is provided by a cathode-ray tube at a frequency of 50 Hz. The light from the image is passed via a condenser lens to a photo-multiplier cell. A range of the harmonics of the signal from the photo-cell is selected by means of a band-pass filter. The instantaneous amplitude is squared by a nonlinear circuit and then averaged by a low-pass filter. The focusing motor is controlled by a peak-holding type of controller followed by a bistable circuit which changes state and reverses the drive motor each time that a signal is received from the peak-holding circuit.

Such work can probably be adapted directly for use with the

robot eye, though here the scanning can make use of the many photo-cells in the eye rather than a special scanning illuminant.

This method of focusing by moving a rigid lens is in fact that used in the fish. In the human the actual shape of the flexible lens is changed.

It has been suggested that, in focusing the eye, the human control system operates purely on magnitude information ('even-error signal') and not in addition on sign information ('odd-error signal'). If the image on the retina is blurred, then the initial trial-and-error focusing movement of the eye can easily be in the wrong direction.[10]

It would be desirable if this human fault could be overcome in the robot. The necessary information about sign could be obtained by superposition into the focusing action of an additional small oscillation at, say, 2 Hz. Such an oscillation is indeed known to be present in the human focusing system, yet investigations indicate that this action does not provide any sign information, since the initial movement can still be in the wrong direction.

Perhaps binocular vision could be of assistance here. An alternative possibility might be the detection of chromatic aberration in order to obtain a sign signal for correction.

A promising form of lens[19] for robot use is that having one flexible wall, the curvature of which is altered by pumping in a liquid. A mixture of saturated aqueous solution of calcium bromide with glycerol is used, to obtain a refractive index of about 1.5, the same as that for glass. The actuating cylinder is 3 mm in diameter and 20 mm long. The lens consists of three layers: there is a front glass lens; then a layer of polyvinyl butyral which adheres to the glass and has a hole in the centre; then a layer of thin coverslip glass which is flexible. Distant vision uses the whole lens, but near vision is restricted to the centre.

In one form of camera focusing device[25] a spot of infra-red light, chopped on and off at between 20 and 70 Hz, is projected on to the object to be viewed. The mechanism then focuses the reflected image of this infra-red spot so that at the same time it automatically focuses the visible light reflected from the object. The light spot is about 15 cm in diameter at a distance of about 20 m from the camera. A technique such as this, which effectively selects and separates out a very small portion of the scene to be viewed as being at the correct distance, appears to be very promising for application to the focusing of the robot eye.

As many as 23 miniature electric motors are used to give automatic and remote control of some colour TV cameras, by the operation of precision multi-turn potentiometers[26].

Adaptive focusing of a television camera[9, 11]

Where a TV camera tube is used as a robot eye system, it is necessary to incorporate automatic focusing. Various approaches to this are possible, provided that the object to be looked at lies all in one plane. Where this is not so, there is the additional problem of the decision as to which object in the field of view should be focussed, unless a pinhole system is used.

One approach to the focusing of a TV camera which has been developed at Aston by R. J. Davies has used a perturbation technique[31]. In order to avoid the necessity for rapid movement of the tube or lenses, the perturbation is obtained by rotation of discs of Celastroid in front of the camera tube at a speed of 1500 rev/min, giving a perturbation frequency of 25 Hz. Disc thicknesses of 0.5 mm were used for rigidity, although these gave slight picture degredation, especially when bending appeared after some use. A Velodyne was used to move the Vidicon camera tube for focusing, through a train of gears and a screwed rod. Since the total travel of the tube is only just over 1 cm, a fine (10 B.A.) thread is used on the screw drive.

This approach has given a good measure of success. However, much further work is required in order to achieve totally satisfactory systems, and this might not be justifiable until there has been further work on the central control of attention. At the same time, the work has been so successful that it is clear that the method apparently used by humans and animals of perturbation, whether oscillatory or random, will probably be the best for use in robots.

Persistence of vision

There is one form of optical illusion which is so important in the modern world that it is taken for granted. This is the phenomenon of persistence of vision. If human vision was not subject to the persistence effect, then devices such as films and television would not work.

Now, in the case of a robot, persistence of vision is quite optional, and in fact it might prove difficult to match the human characteristics under all conditions of lighting, etc. Consequently, a decision has to be taken at the design stage with any robot fitted with visual sensors as to whether persistence should be incorporated, and if so what should be the period of the persistence.

It is probably best, for general-purpose use, if the length of inherent persistence in the robot visual sensors is maintained as short as possible. It is then possible to incorporate an artificial persistence of any required length at some later point in the visual system. However, the size, weight and cost of the extra provision must be balanced against the advantage of standardisation of the primary sensor.

There are some applications where the slowness of reaction of the human to a visual stimulus is a disadvantage which can be overcome by the use of a robot having a rapid visual reaction. Such a robot could observe, and act on, high-speed reactions in mechanical or in chemical systems, for example.

A useful device for short-term visual memory[21] has recently been produced by R.C.A.[12] This is a silicon storage Vidicon tube, which has a target formed from a silicon wafer containing 600 000 photodiodes, each isolated by silicon dioxide insulation. The diodes are scanned by a low-velocity beam. A high-velocity beam is used for erasure of previously recorded material; a lower-velocity beam is used for storage of information on to the silicon diode matrix; and the lowest-velocity beam is used for reading the stored information. A gradual erasure of stored information is caused by the read beam, though a long-term storage is possible if there is no reading. This device would seem to be ideal for short-term storage of memory in the robot, especially since it is much smaller and lighter than is magnetic tape or drum recording equipment.

Miniature radars and Doppler effect[20]

The introduction of the Gunn-effect diode as a generator of microwave oscillations has brought about the possibility of an extreme of miniaturisation in radar equipment, to the point where it becomes of interest in robot applications. As an example, the Scott Mini-Radar measures only $12 \times 9.5 \times 5.7$ cm, not including the horn antenna, and it operates on a supply of 12 V d.c. at 0.3 A. The weight is only 860 g. It normally operates at a frequency of 13.418 GHz, with a power output of 5 mW. The radar is usable for detection ranges of, for example, 15 m for a walking human to 125 m for an object such as a bus. With a parabolic dish antenna a beam width as low as 5° can be obtained.

Thus it can be seen that, despite the portability of the equipment, quite a useful range of operations can be obtained. However, such a device could not be fitted as standard at present on a general-purpose robot because of the cost. A lower-cost equipment to the same basic design has more recently been introduced for Doppler applications. It becomes clear that if such equipment was required in very large quantities for robot applications, the cost would become very small.

240 Perception of Movement

The advantage of miniature radar devices over visual devices for use in the robot is that for simple obstacle detection not only is ranging of the obstacle relatively easy, but in addition there will be less extraneous interfering information.

The use of lasers with mirror scanning has been proposed for obstacle detection on a mobile robot device for use on the surface of Mars[24]. Automatic control of the throttle and brakes of a car in order to prevent the possibility of nose-to-tail collisions has been produced experimentally. Radar was used to measure the inter-car distance, while the speed of the car being controlled was measured from the propshaft[27]. Light-emitting diodes have also been used[33].

There have been many attempts to produce electronic devices which can be used by blind people for the detection of obstacles when they are walking along[28]. It is essential that such a device be both small and of light weight, and these are just the requirements for equipment to be used on a robot[34].

It appears to be possible that certain birds can detect the magnetic field of the earth and that a locally affixed magnet can disrupt their direction finding[29]. Whether or not this is verified, there is no reason why a robot should not have a built-in sensor of the earth's field for direction sensing, though this would naturally be unreliable in certain robot environments.

REFERENCES

1. Zusne, L., *Visual Perception of Form*, Academic Press (1970).
2. Ballantine, J. M. and Allan, W. B., 'Fibre Optics', *Sci. J.*, **1**, September, 1 (1965).
3. Robinson, D. A., 'A Method of Measuring Eye Movement Using a Scleral Search Coil in a Magnetic Field', *Trans. IEEE*, **BME10**, October, 137 (1963).
4. Comeau, C. P. and Bryan, J. S., 'Headsight Television System Provides Surveillance', *Electronics*, **34**, November 10, 86 (1961).
5. Goertz, R. G., *et al.*, 'An Experimental Head-Controlled TV System to Provide Viewing for a Manipulator Operator', *Proc. 13th Conf. Remote Systems Technol., 1965* (A.N.S.), 57.
6. Goertz, R. C., *et al.*, 'The ANL Mark TV2—An Experimental 5-Motion Head-Controlled TV System', *Proc. 14th Conf. Remote Systems Technol., 1966* (A.N.S.), 124.
7. Walter, W. G., 'The Past and Future of Cybernetics' in: Rose, J. (ed.) *Human Development, Progress of Cybernetics*, 45, Gordon and Breach (1970).
8. Grindley, G. C. and Townsend, V., 'Visual Search without Eye Movement', *Q. J. Exp. Psychol.*, **22**, 62 (1970).
9. Deeley, E. M. and Allos, J. E., 'Automatic Focusing of an Optical System by Extremum Control', *Proc. IEE*, **114**, January, 161 (1967).
10. Stark, L. and Takahashi, Y., 'Absence of Odd Error Signal Mechanism in Human Accommodation', *Trans. IEEE*, **BME12**, July, 138 (1965).
11. Curling, C. D., *et al.*, 'Focusing Aid for an Electron Microscope', *Proc. IEE*, **116**, March, 334 (1969).

12. Anon., 'Now a Stop Action Camera Tube Design', *Electronics*, **8**, January, 13 (1971).
13. Stark, L., et al., 'Nonlinear Servoanalysis of Human Lens Accomodation', *Trans. IEEE*, **SSC1**, November, 75 (1965).
14. Roberts, L. G., 'Picture Coding Using Pseudo Random Noise', *Trans. IRE*, **IT8**, February, 145 (1962).
15. Limb, J. O., 'Design of Dither Waveforms for Quantized Visual Signals', *B.S.T.Jl*, **48**, September, 2555 (1969).
16. Lippel, B., et al., 'Ordered Dither Patterns for Coarse Quantization of Pictures', *Proc. IEEE*, **59**, March, 429 (1971).
17. Evans, C. R., 'Fragmentation of Patterns Occurring with Tachistoscopic Presentation', *IEE/NPL Conf., Pattern Recognition, 1968*, 250.
18. Rees, M. G., 'An Artificial Eye for use in Automatic Handling Systems', Ph.D. Thesis, Engng Dept., University of Cambridge (1968).
19. Anon., 'Variable-Focus Lenses', *Control*, **12**, May, 469 (1968).
20. Anon., 'Mini Radar Runs a Model Railway', *New Scient.*, **29**, March 31, 835 (1966).
21. Averbach, E. and Coreill, A. S., 'Short Term Memory Vision', *B.S.T.Jl*, **40**, 309 (1961).
22. Ratcliff, F. and Riggs, L. A., 'Involuntary Motions of the Eye During Monocular Fixation', *J. Exp. Psychol.*, **40**, December, 687 (1950).
23. Dreyfus, M. G., 'Space Age Production by Automatic Image Alignment', *Mfg Eng. Mgmt*, **66**, March, 29 (1971).
24. Kuriger, W. L., 'A Proposed Obstacle Sensor for a Mars Rover', *J. Spacecraft Rockets*, **8**, October, 1043 (1971).
25. Odone, G., 'Camera's Infra Red Eye Focuses on New Vistas for Ranging', *Electronics*, **43**, April 27, 10 (1970).
26. Anon., 'Micromotors Used in Automatic Camera Alignment', *Electron*, May 4, 22 (1972).
27. Anon., 'Stopping Those Nose-to-Tail Pile-ups', *Electron*, September 14, 8 (1972).
28. Beurle, R. L., 'Electronic Guiding Aids for Blind People', *Electron. Eng.*, **23**, January, 2 (1951).
29. Southern, W. E., 'Magnets Disrupt the Orientation of Juvenile Ring Billed Gulls', *Bioscience*, **22**, August, 476 (1972).
30. Koenderink, J. J., et al., 'Foveal Information Processing at Photoptic Luminances', *Kybernetik*, **8**, April, 128 (1971).
31. Davies, R. J., 'Adaptive Focussing of a Television Camera', University of Aston (1968).
32. Sugie, N. and Jones, G. M., 'A Model of Eye Movements Induced by Head Rotation, *Trans. IEEE*, **SMC1**, July, 251 (1971).
33. Mims, F.M., Use L.E.D.s, Not Lasers, In Rangefinders, *Electron. Des.*, **20**, May 25, 48 (1972).
34. Gilder, J. H., 'Space-Age Technology Opening New Doors for the Blind, Deaf and Crippled', *Electron. Des.*, **20**, May 25, 24 (1972).
35. Heginbotham, W. B., et al., 'The Nottingham SIRCH Assembly Robot', *Proc. 1st Conf. Industrial Robot Technology, Nottingham, March, 1973*, 129.

14

Human hearing

The robot ear

Robot devices which could hear the voice of man, and reply to it, have been sought for since the earliest days of civilisation[1]. In the Middle Ages reports of hearing and speaking robot devices became well documented[22]. However, it is difficult to decide how much of these reports is fact and how much is as fanciful as the legendary door which responded to the command of 'Open Sesame!'

We can here disregard such early reports and ignore such early trickery as that of von Kempelen[23] and directly consider more recent and more practical work. It is not difficult to construct a mechanism which will respond only to a single frequency in speech, such as, for example, the early speech-operated dog 'Radio Rex' described by Paget[42]. However, the problem of analysing and recognising human speech is much more intractable.

The human ear contains a number of resonators known as hair cells, arranged along a vibrating membrane known as the basilar membrane. There are about 30 000 of these hair cells, arranged in four rows, in the human ear. Tiny hairs grow from each hair cell, and these are used to translate the movements of the membrane into electrical signals produced by the cell. The electrical signals are taken via nerve fibres from the ear to the brain.

The exact nature of the mechanism of perception of sound by humans and by animals is not well understood, despite extensive research and even more extensive publication. Fortunately, the cybernetic engineer does not have to attempt to reproduce the exact form of the human or animal ear. It is sufficient to produce a working device for robot use. However, the cybernetic engineer always hopes that his work will help to stimulate fruitful thought in the physiologists and psychologists who are engaged on investigations into the animal hearing mechanism.

Investigations of the characteristics of the human ear[2] indicate that it behaves rather like the equivalent arrangement shown in the block diagram of *Figure 14.1*. The middle ear acts as an effective low-pass filter, having a cut-off frequency of about 1500 Hz above which the response falls off at the rate of 18 dB/octave. The resonators arranged along the basilar membrane act much as band-pass filters tapped on to a delay line. Each of the band-pass filters acts effectively as a tuned circuit having a Q value of only about 1.5, so the resonators are not highly selective.

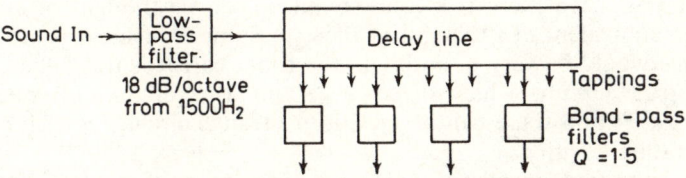

Figure 14.1 Block diagram showing characteristics of the human ear

The ear appears to signal the three features frequency, intensity and time to the brain. In addition, there is the valuable binaural information produced as a result of the possession of two separated ears.

There has been a great deal of work on speech communication systems based on the frequency analysis capability of the human ear. This has been largely based on the work on the Vocoder[19], first described by Dudley[24]. The rate of variation of acoustical energy at any given frequency contained in speech is relatively slow, so that the analysis of speech into its constituent frequencies, the transmission separately of the amount of energy in each of these frequencies and the reconstitution of the original speech from the energy information appears to be a promising process for speech communication.

Unfortunately, the process has never achieved acceptance, probably because of the impersonal nature of the reconstituted speech. In the case of the robot hearing system, we do not have to worry about the 'naturalness' of reconstitution, since we are only concerned with the possibility of recognition of speech sounds by the robot. As an example, the robot must be able to accept and to act upon verbal orders received from a human.

Although there are many nerve cells in the human ear, it is only possible for the human to distinguish tone intervals in which the change of frequency δf is related to the base frequency f by the relationship

$$\frac{\delta f}{f} = 0.033$$

This indicates that only about 60 separate tones can be distinguished by a human in the speech range 300–3000 Hz. There would seem to be little use in providing a robot with a greater capability than this, and probably a much reduced capability will suffice.

In the human ear a form of reflex protection against the damaging effects of loud noises is provided by the stapedius muscle[3]. This contracts when there is a loud sound. It can be thought of as the aural equivalent of the pupil of the eye. Some equivalent protective device would be very desirable in the robot ear. It is not difficult to construct circuits which detect a certain level of audio-frequency energy[4, 26] and these can be used in protective circuits as well as in activating circuits.

A suggested construction for an array of passive filter circuits which simulate the frequency response of the human cochlea has been presented by Bolie[21]. Each section consists of a series arm of a 100 Ω resistor in series with a certain value of inductance L H, followed by a shunt arm formed from a 100 ohm resistor in series with a capacitance C and the same value of inductance L as used in the adjacent series arm. The resonant frequencies of the shunt arms decrease exponentially with the increase of the section number along the cochlea. A total of 80 sections is used, terminated by a 1000 Ω resistor.

The response of the human ear to sound is not constant but depends on the nature of the sound being heard[37]. For example, not only does the human ear not respond uniformly to all frequencies of sound, but in addition the actual shape of the response–frequency curve obtained varies in accordance with the loudness of the sounds[38, 39]. It will probably not be worth while trying to reproduce exactly such characteristics of human hearing for application to the robot, and it is quite possible that we shall encounter some robot aural illusions which are quite different from the aural illusions experienced by the human.

Work on speech recognition

The writer has reviewed elsewhere[1] the extensive work being carried out in the field of speech recognition. Suffice it to mention here that most of the work involves machines which have to be set up by a human rather than machines which are themselves inherently capable of learning.

It seems to have been widely overlooked that without some form of feedback, which produces a learning loop, a recognition accuracy of 100% is basically required. However, with the addition of verbal and visual (and perhaps tactile) feedback from the listening robot, indicating to the speaker the recognition or understanding or otherwise, a much lower basic standard of accuracy is acceptable. In addition, the more continuous the feedback the greater is the over-all speed of communication.

The so-called motor theory suggests that all human learning of speech sound perception is based on reference to the articulatory movements which would be required for the human to produce that sound. Such a theory does not help to explain how a human can learn to recognise sounds which he is quite incapable of producing verbally, and it does not appear to be a very fruitful approach for the robot learning of speech recognition. However, if it is true, it might help to overcome some of the problems of human perception of speech.

It is important to note that the speaker-to-speaker variation of speech sounds is very great, and yet this does not prevent the human from perceiving and understanding the meaning of speech. In some cases a period of conditioning is necessary before the speech can be understood: for example, when the speech is in a very strong and unfamiliar accent. However, even in such cases the human rapidly learns to adapt to the speaker.

There is a very great variation in the nature of the speech from a single human being under certain circumstances. For example, when the human is tired, or perhaps drunk, the speech becomes slurred and so less clear. Another circumstance in which there is a great variation of the nature of the speech from a single human is when he is under great stress. A typical case of this might be the air traffic controller, and a robot equipment designed to work on air traffic control in conjunction with a human would have to be capable of recognising such variations and of making allowance for them. It has been suggested that such stress variations are actually greater than many speaker-to-speaker variations.

An extremely important application for a robot which could recognise speech sounds would be as an automatic language translator. Sadly, work on the automatic translation of languages using computers has so far served mainly to reveal the many difficulties. Among the many of these perhaps the simplest is that of the *double entendre*. Probably not until we can build complex learning robots (with feedback of results) will competent automatic translation be achieved. Progress in the field of speech recognition[18, 20] is extremely slow, though there are some signs of progress towards the goal of recognising continuous connected speech. Hill[33] has reviewed the field and reported on the work carried out recently, and he suggests that real progress is at last being made in research on the recognition of connected speech. For example, based on the earlier work of Reddy[34, 35], the scheme developed by Vicens is capable of dealing with connected speech commands to a robot arm[36] and to a desk calculator. In this work it has been recognised that one significant problem is that of providing for rapid recovery after a complete failure of recognition.

The nature of speech[5-8]

Dudley has suggested that human speech can be regarded as consisting of two modulated carriers. One of these carriers is the 'buzz' of the vocal cords; the other is the 'hiss' caused by the escape of breath. The carriers are then modulated by slow variations of resonance, pitch and intensity, in order to convey information.

From the human ear, information about the sound waves heard is passed to the brain in terms of frequency, intensity and time. It seems that the brain is capable of interpreting meaning simply by use of these signals.

The instantaneous peaks of the speech waveform of voltage from a microphone exceed the root mean square value by 12–15 dB. In addition, the average vowel produces a peak microphone voltage about 12 bB higher than that produced by the average consonant, though again there is a large variation of the instantaneous ratio. It is therefore rather surprising that the intelligibility of speech appears to depend much more heavily upon the sounding of consonants than upon vowels.

Now if the peaks are clipped from a speech waveform, both the ratio of peak to r.m.s. and the ratio of vowel to consonant amplitude are reduced simultaneously. This increases the relative weighting of consonants, and there should therefore be a possibility that such a clipping procedure will increase intelligibility of speech.

Unfortunately, the quality of severely clipped sound waves is sometimes rather disturbing to an unaccustomed human listener. However, this should not prevent the use of such principles with robots, where the subjective effect is of no importance when intelligibility is considered. It is perhaps worth mentioning that peak clipping has actually been found to give a very useful increase of intelligibility with human listeners, particularly under very noisy conditions, even though the human listener does not 'like' the sound.

There is one major disadvantage of severe amplitude limitation which can be most important in its application to the mobile robot. High-gain amplifier circuits are required to follow the clipping process in order to increase the very low amplitude of the clipped voltage waveform up to a usable level. Such high-gain amplifiers are very susceptible to interference effects caused by pick-up of stray signals, and this fact can be very troublesome in a very variable environment such as that of a mobile robot.

Information content of human speech[9, 10]

It is of interest to consider human speech communication requirements from the information theory point of view[25], since this can help to suggest the requirements for robot systems.

The writer has pointed out elsewhere[25] that in order to communicate human speech with a frequency bandwidth of 5000 Hz, under conditions where the signal-to-noise power ratio is 1000 (or 30 dB), the capacity of the communication channel must be 50 000 bits/s. Although this capacity is required for the communication of normal speech, it is known that the human brain is quite incapable of decoding or of making use of information at such a rate. Flanagan[8] has reviewed a number of attempts at the estimation of this 'syllabic' rate, and from these investigations it is suggested that the human brain is not capable of handling information at a greater rate than about 50 bits/s. The actual maximum achievable rate is probably even less than this. For example, a human operator receiving morse code signals at about 35 words/min is only achieving a syllabic rate of about 6 bits/s.

Under special circumstances, however, the human brain can certainly handle greater rates. Jacobson has suggested that the information capacity of the human ear can be as high as 8000 bits/s, corresponding to an average capacity of about 0.3 bits/s for each nerve fibre coming from the cochlea.

A human can understand speech at a rate of several hundred words per minute. Consequently the rate of perception of phonemes, or bits of speech, is about 20/s. It has been suggested by Cooper[11] that the disparity between this high rate, as compared with the auditory processing capability of only about 6 bits/s, is explained by a process of parallel encoding of phonemes.

If the human is indeed incapable of handling information rates greater than about 50 bits/s, then there is certainly no justification in attempting to construct a general-purpose robot with a greater capability. However, special-purpose robots which are capable of handling audible information at rates greater than those which can be achieved by a human will certainly be useful in some applications.

Limitation of frequency bandwidth

Experiments with human listeners on the limitation of the audio-frequency band have indicated that there is little loss of intelligibility if all speech frequencies below about 350 Hz are removed. Indeed, in good listening conditions little more is lost if the lower cut-off frequency is increased to 580 Hz.

At the high-frequency end of the speech frequency spectrum it is found that the intelligibility to a human listener is very little reduced if frequencies above about 3900 Hz are removed. However, the intelligibility is seriously impaired if this upper cut-off frequency is reduced below 2500 Hz. It should perhaps be mentioned that figures such as these have been obtained using human listeners and filters having an extremely sharp rate of cut-off.

One point of interest is that the peripheral form of sound frequency analysis performed in humans by the basilar membrane and the hair cells is not present at all in fishes and lower forms of life. Here it appears that nervous pulse signals having the actual frequency of the applied sound are taken directly to the brain for analysis.

The single equivalent formant[12]

It is usual to consider speech sounds as consisting of the combination of three components each of which is at a different formant frequency. It is possible to replace these three formants by a single equivalent formant frequency while still retaining the phonetic content of the speech sound.

It has been proposed that it is possible to replace the many slowly varying outputs from a conventional Vocoder form of system by only three parameters, all varying only slowly. The three parameters are:

1. Single equivalent formant frequency.
2. Single equivalent formant amplitude.
3. The amount of 'voicing' in the sound.

In order to simplify the extraction of the frequency and amplitude

Human Hearing 249

of the single equivalent formant, it has been found possible to operate on the length and amplitude of the first half-cycle of speech sound signal following either a glottal stop or a rapid rise in the waveform of an unvoiced sound or fricative.

Although work on such a system has used fairly complex circuitry, it has at least shown promise of giving a measure of speaker invariance. It is hoped that further work on this approach may give a simplified solution to the problem of speech recognition.

The structure of spoken words

For the recognition only of spoken numerals, together with a few words such as 'plus', 'minus', 'total', it has been found possible to operate simply on two different types of consonant sound. These are the explosives (s, t, etc.) and the soft sounds (th, f, etc.). Vowel sounds are required in addition, the relative phases of the components of the sound here being used for recognition.

In this approach, three different parameters are recognised in each of three different time positions, the time order being consonant–vowel–consonant.

In a direct human conversation a form of visual feedback is provided by the nods and by the expression on the face of the listener, and this feedback provides an aid to communication. Telephone conversation suffers from the lack of such direct continuous feedback, and this fact can help to slow down the rate of telephonic communication.

Visual feedback is not present in the case of human-to-robot conversation, and it has consequently been found useful to provide a form of audible feedback such as an audio 'pip' for acceptance or understanding or a 'buzz' for non-acceptance or non-recognition. A visual feedback from a lamp could easily be provided simultaneously or alternatively.

At the end of a sentence, in one form of speech recognition equipment, the human can ask verbally for the sentence to be repeated back by saying 'check', and he can then confirm verbally 'right' or 'wrong'. While this slows down the rate of human-to-robot communications considerably, it is important to maintain the rate of error as low as possible in early robot applications in order to encourage acceptance. It is better that a robot receiving verbal instruction should ask for a repeat of the instructions rather than run any risk of carrying out the wrong operation.

The classification of the sounds heard by man is performed on the basis of a vast store of previous experience which has built up the knowledge of the language system which is possessed by the human. Now because of the high redundancy of human speech, it is possible for the listener to give a much greater weight to his store of information about the language than he gives to the actual instantaneous acoustic information. In this way a man can make sense of running speech.

Most speech recognition machines have relied entirely on the incoming information about the waveform. This has been a fairly successful approach where only a limited vocabulary has to be recognised, particularly if no more than the speech sounds produced by one individual human are involved[47].

However, this approach cannot be expected to deal successfully with even a limited form of running speech. Here the only possible approach is to build a machine which is capable of learning, and not to try to set it up according to our own preconceived ideas[31]. The machine should be provided with a form of cochlea and left to associate the sounds heard[47]. The process will be slow, as with a human.

Miscellaneous points

One very interesting feature of human hearing is known as the 'cocktail party effect'. In a room full of people, all talking at once, it is possible for a human listener to concentrate on one speaker and to ignore all of the other sound. This form of concentration is carried out in the central nervous system, and it is not known at present how such a feature could be incorporated into a robot hearing system.

A human can recognise the voice of the person speaking, if it has been heard before. This is probably done on the basis of a frequency analysis. It is possible to recognise the person from a printed running spectrum of frequency against time, and there have been serious suggestions that such 'voiceprints' should be used as evidence in a court of law[13-15]. However, this would be most undesirable, because a good human mimic can produce a spectrum almost indistinguishable from that of the person being imitated[43]. Ladefoged has been quoted[30] as saying in connection with work on speaker identification that: 'You really want to know if it's the President saying "Drop the bomb now!"' One trembles to think that perhaps such reliance is being placed on the identification of a voice on a telephone, and that the whole future existence of mankind might be prejudiced by simply ignoring the alternative possibility of having a system of complex coded messages. However, perhaps this can explain the heavy expenditure on investigations into speaker identification! It would be desirable if a robot could 'recognise' the voice of the person who is giving it orders[48], but such a development is unlikely in the near future.

In the human, feedback of speech to the ear can be both internal and external. Internal feedback is via bone conduction of sound. External feedback is via the ears. The latter is most important, and alteration of the feedback path—for example, by the introduction of certain delay times—can have a completely disruptive effect on the production of speech by the human. The same would be true of a robot.

There will be little difficulty in the construction of robots having ultrasonic, or indeed subsonic, hearing capabilities. For certain industrial applications such robots will be extremely useful in that they will be able to perform functions which could not be carried out by a human.

One point of importance is that, in order to perceive the meaning of speech, it is necessary to take into account the order in which the various speech sound are produced. Consequently, each decision about meaning is based on a previous decision, and there is a running segmentation of the speech by the listener. Now if decisions are based purely on the order of the speech segments, there is no need to store or to act on the length of the speech segment. This is an over-simplified view, however, since sometimes the length of a sound can convey meaning.

One field in which there might well be mutual exchange between the cybernetic engineer and the medical engineer is the production of the artificial larynx. There have long been experiments on the production of such devices for medical use, though these have, in general, been limited to a very simple approach.

An important point which is often overlooked when speech recognition is considered is the fact that the rhythm alone of sounds can convey a great deal of information. Consider, for example, how the rhythm of a railway train on jointed track seems to 'speak' sometimes, and how music can be imagined in a noisy but rhythmic background of sound.

Audible illusions[16, 17]

In the same way that optical illusions are experienced by the human eye, audible illusions are experienced by the human ear. The nature of these illusions might help to give us a clue to some of the features of learning in the human hearing system, and indeed in the human nervous system as a whole. However, as mentioned in connection with optical illusions, it is necessary to exercise caution in making use of data on illusions, unless one is quite sure that the illusion in question is a truly universal feature of all humans, regardless of race or background.

One well-known audible illusion is produced by monotonous repetition of a word. It is quite easy to carry out such an experiment by making a loop of magnetic tape upon which a word is recorded, and then replaying the loop repeatedly for several minutes. As an example of such an experiment, Warren[16] quotes the case of repetition of the word 'tress'. It was heard as, for example, 'stress', 'dress', 'Jewish', 'Joyce', 'Jewess', 'floris', 'florist' and 'purse'. Warren has compared this process with the breaking-up of a visual image which occurs in the case of retinal stabilisation. He suggests that the process is in two parts: first the loss of meaning, then the mental reorganisation of the stimuli into a meaningful form.

It seems that perhaps the brain is not well adapted to the determination of either temporal or spatial order in which stimuli are presented, unless there are other clues. This is a most significant point.

Binaural hearing

The two ears of man and of animals are used for two purposes. Firstly, there is the obvious one of redundancy. Hearing is a most important sense, and it is important that if the use of one ear is lost, either temporarily or permanently, then the other ear is still usable. However, there is an important use of the two ears simultaneously which is of great assistance in everyday life. This is the localisation of sounds which can be carried out in the brain by making use of differences between the signals from the two ears.

Today this localisation is utilised in the system of stereophonic sound reproduction of signals from radio and from gramophone records. However, it has more vital uses. An animal which can use its binaural hearing in order to avoid trouble increases its chances of survival, and so this sense is favoured in natural evolution.

However, is it worth adding complication to a robot in order to give it this directional hearing? For the general-purpose robot it is probably not worth the additional circuit complication which is necessary, though the sense might be valuable in certain special-purpose robots.

If a source of random noise is used, a human listener judges direction by a comparison of the times at which the sound arrives at the two different ears, or so it appears from human experiments. If the difference between the times of arrival at the two ears is small, say less than 1 ms, the direction sensitivity is very poor for random noise.

However, if a pure tone (for example, of 1000 Hz) is heard, then the judgement of direction is directly dependent on the relative time displacement between the signals at the two ears. If the displacement in time of the two signals is varied artificially and continuously, the direction of the source which is guessed by the listener changes over from one side to the other quite sharply. Such tests can be carried out with a human observer by applying signals having the same shape but a different time of arrival to the two ears—for example, via two different earphones.

Certain forms of speech sound have been shown to be localised in the same way by humans, and it has been suggested that there is some form of cross-correlation carried out in the nervous system in order to facilitate this action. Thus in any robot in which such binaural localisation is required it will, it appears, be necessary to carry forward to the central nervous system some information about the relative time of arrival or of the relative phase of the signals arriving at the two ears.

In some forms of radar equipment, sometimes using visible or infra-red light, the relative strengths of the signals arriving at two sensors have been used for direction sensitivity. If a single scanning sensor is used, the source direction is indicated by equality between the strengths (and possibly also the phases) of the two signals arriving at two different points on the mechanical scan. A simpler alternative, where a mechanical scan is used, is to determine the direction at which a maximum strength of signal is received by a highly directional receiver, using for example, a parabolic scanning reflector.

The actual method to be adopted in any particular case depends on the task which must be carried out by the robot, and it is very unlikely that binaural hearing will be required on early general-purpose robots.

If two (or more) separate ears are fitted to a robot, the directionality which is thereby achieved can be utilised to control the direction in which the whole robot faces or the direction in which the robot head is turned. In such cases it is, of course, necessary that the output from the ears should be taken to the corresponding movement control system, and this is likely to involve problems of servo stabilisation which again make its use unlikely on early general-purpose robots. The front-to-back directivity feature introduced by the shape of the external ear is very easy to add to the robot ear, and a wide range of different polar diagrams of directional sensitivity is possible.

In some cases it will be advantageous to fit the ears of the robot not in the head but in the body or, for some purposes, even in the hand. The radar-like obstacle-avoiding ability of the bat, using ultrasonically produced and perceived sound, might be very useful in some forms of robot.

Microphones[27, 28, 32, 40, 41]

It is unlikely that anything like the animal method of sound detection, involving, in effect, the use of large numbers of individual microphones each tuned to a different frequency, will be used in the robot. Instead a single microphone will be used to feed into electronic filtering circuits.

The form of microphone to be used will be determined by the simplicity and the reliability which is necessary. Although the simple carbon microphone as used in the telephone has attractions, it is very noisy and is affected by mechanical movement. The crystal form of microphone can give a large electrical output, though it is not usually suitable for use at high temperature and its high impedance makes it susceptible to interference in the unpredictable and mobile environment of the robot. Crystal microphones have recently been designed complete with a self-contained integrated amplifier operating from the low-voltage supply which is usually used to supply a carbon microphone.

Other forms of microphone are, in general, too expensive or too fragile or give too low an electrical output. However, although the moving-coil form can only give a low value of electrical output, the fact that it can also be used as a loudspeaker makes it attractive for use in a robot, since it can be used both as ear' and as mouth. A return to the earlier moving-iron device, as used in the telephone earpiece, is not impossible. Whatever form is used, high fidelity is not an urgent requirement for robot use. Crystal microphones are sometimes used as loudspeakers.

Most forms of microphone have a directional form of response, and this can be valuable. For example, if binaural hearing is required in the robot, then the directional feature is very necessary and two separate microphone ears must be fitted.

It is difficult to envisage the use of the noise-cancelling form of microphone in the robot, and instead it will be necessary to make provision for discrimination against interfering signals in the robot nervous system. In effect, some form of 'attention fixing' system is required, though much work will be required before this is achieved. In the mobile robot some form of anti-vibration mounting for the microphones will be essential if the hearing is not to be paralysed whenever the robot moves.

Possible methods for replacement of the carbon granule microphone in telephony are receiving some attention, and if this work is successful it will, of course, be of great interest in the robotic field for use as a robot ear. However, work using electrets[29], which have a high impedance and are humidity-sensitive, does not look too hopeful for robotic applications.

REFERENCES

1. Young, John F., *Cybernetics*, Iliffe (1969).
2. Lummis, R. C., 'The Secret Code of Hearing', *Bell Labs Rec.*, September, 261 (1968).
3. Dallos, P. J., 'Study of the Acoustic Feedback Loop', *Trans. IEEE*, **BME11**, January, 2 (1964).
4. Keen, M. J., 'A Voice-Operated Electronic Switch Having a Low Drift with Temperature', *Electron. Components*, **10**, May, 577 (1969).
5. Allen, V. H., 'Laboratory Equipment for Quantizing Speech', *Electron. Eng.*, **28**, February, 48 (1956).
6. Licklider, J. C. R., 'Effects of Amplitude Distortion upon the Intelligibility of Speech', *J. Acoust. Soc. Am.*, **18**, 429 (1946).
7. Licklider, J. C. R., 'The Intelligibility of Amplitude-Dichotomised Time-Quantised Speech Waves', *J. Acoust. Soc. Am.*, **22**, 820 (1950).
8. Flanagan, J. L., *Speech Analysis, Synthesis and Perception*, Springer-Verlag (1965).
9. Jacobson, H., 'The Information Capacity of the Human Ear', *Science, N.Y.*, **112**, 143 (1950).
10. Jacobson, H., 'Information and the Human Ear', *J. Acoust. Soc. Am.*, **23**, July, 463 (1957).
11. Cooper, F. S., 'Speech—Man's Natural Communication', *IEEE Spectrum*, **4**, June, 79 (1967).
12. Teacher, C. F., *et al.*, 'Experimental Limited Vocabulary Speech Recogniser', *Trans. IEEE*, **AU15**, September, 127 (1967).
13. Bolt, R. H., *et al.*, 'Speaker Identification by Speech Spectrograms: A Scientist's View of Its Reliability for Legal Purposes', *J. Acoust. Soc. Am.*, **47**, February, 597 (1970).
14. Das, J. K. and Mohn, W. S., 'Pattern Recognition in Speaker Verification', *A.F.I.P.S. Conf. Proc.*, **35**, Fall, 721 (1969).
15. Li, K. P., *et al.*, 'Experimental Studies in Speaker Verification Using an Adaptive System', *J. Acoust. Soc. Am.*, **40**, No. 5, 966 (1966).
16. Warren, R. M. and Warren, R. P., 'Auditory Illusions and Confusions', *Sci. Am.*, **223**, December, 30 (1970).
17. Ptacek, P. H. and Pinheiro, M. L., 'Pattern Reversal in Auditory Perception', *J.A.S.A.*, **49**, February, 493 (1971).
18. Newell, P. H. and Barr, J. S., 'Voice Operated Powered Devices', *Mech. Eng.*, **92**, November, 25 (1970).
19. David, E. E., 'The Present Status of Vocoder Speech Bandwidth Reduction Systems', *Proc. Natl Electron. Conf.*, **19**, 730 (1963).
20. Lavoie, F. J., 'Voice Actuated Controls', *Mach. Des.*, **42**, January 22, 135 (1970).
21. Bolie, V. W., 'Computer Aided Refinements in the Design of an Artificial Cochlea', *Proc. 22nd S. W. Conf. IEEE, Dallas, 1970*, 308.
22. Cohen, J., *Human Robots in Myth and Science*, Allen and Unwin (1966).

23. Dudley, H. and Tarnoczy, T. H., 'The Speaking Machine of Wolfgang von Kempelen', *J. Acoust. Soc. Am.*, **22**, 151 (1951).
24. Dudley, H., 'Remaking Speech', *J. Acoust. Soc. Am.*, **11**, 169 (1939).
25. Young, John F., *Information Theory*, Butterworths (1970).
26. Anon., 'Voice-Operated Switching for Loudspeaking Telephones', *G.E.C.Jl*, April, 102 (1957).
27. Robertson, A. E., *Microphones*, 2nd edn., Iliffe (1963).
28. Anon., *Microphones* (B.B.C. Eng. Trg. Manual), Iliffe (1951).
29. Reedyk, C. W., 'Electret Transducers Applied to the Telephone', *Trans. IEEE*, **AU19**, March, 1 (1971).
30. Wood, N., 'Toward Speech Processing Machines', *Mach. Des.*, **42**, July 9, 44 (1970).
31. Lamotte, M., et al., 'Simulation de la Reconnaissance des Formes Vocales par Apprentissage', *C. R. Acad. Sci.*, **269A**, 286 (1969).
32. Gayford, M. L., *Electroacoustics*, Butterworths (1970).
33. Hill, D. R., 'Man–Machine Interaction Using Speech', *Adv. Comput.*, **11**, 165 (1971).
34. Reddy, D. R., 'Computer Recognition of Connected Speech', *J. Acoust. Soc. Am.*, **42**, 329 (1967).
35. Reddy, D. R. and Vicens, P. J., 'A Procedure for the Segmentation of Connected Speech', *J. Audio Eng. Soc.*, **16**, 404 (1968).
36. McCarthy, J., et al., 'A Computer with Hands, Eyes and Ears', *Proc. Fall Joint Computer Conf., San Francisco, 1968*, 329.
37. Masterton, B., et al., 'The Evolution of Human Hearing', *J. Acoust. Soc. Am.*, **45**, 966 (1969).
38. Fletcher, H. and Munson, W. A., 'Loudness, Its Definition, Measurement and Calculation', *B.S.T.Jl*, **12**, 377 (1933).
39. Fletcher, H., *Speech and Hearing in Communication*, Van Nostrand (1953).
40. Gayford, M. L., 'High Quality Microphones', *Proc. IEE*, **106B**, November, 501 (1959).
41. Amos, S. W. and Brocker, F. C., 'Microphones', *Electron. Eng.*, **18**, August, 255 (1946).
42. Paget, R., *Human Speech*, 79, Harcourt Brace (1930).
43. Lummis, R. C., 'Speaker Verification: A Step Toward the "Checkless" Society', *Bell Labs Rec.*, September, 245 (1972).
44. Vysotskiy, G. Y., et al., 'An Experiment in Oral Control of a Computer', *Eng. Cybern.*, **8**, March/April, 320 (1970).
45. Von Keller, T. G., 'An On-Line Recognition System for Spoken Digits', *J. Acoust. Soc. Am.*, **49**, pt 2, April, 1288 (1971).
46. Gonschorek, J. and Hinrichs, O., 'A Speech Recognition System with Adaptation to Speech Rate and Volume', *NTZ Commun. J.*, **24**, April, 177 (1971).
47. Haton, J. P. and Lamotte, M., 'Preprocessing and Recognition of Speech: Simulation and Practical Realisation', *Automatisme*, **17**, March, 63 (1972).
48. Atal, B.S., 'Automatic Speaker Recognition Based on Pitch Contours', *J.A.S.A.*, **52**, December, 1687 (1972).
49. Hornsby, T. G., Voice Response Systems, *Modern Data*, **5**, November, 46 (1972).

15

Robot hearing and speech

Machines to recognise human language[20-22, 24-26, 28-33, 36, 37, 41]

A great deal is known about the nature of human language, and it might therefore be thought that it would not be difficult to construct machines which could recognise individual words. Indeed, there has been much work in this field. However, success still seems to be as elusive as ever. While it has been possible to build machines which could be set up by a human to recognise words spoken by a single other human, it has proved far more difficult to set up any machine in such a way that it can recognise a number of words irrespective of the speaker or of the accent.

Attempts to construct machines which can recognise human speech have been reviewed by the writer elsewhere[27]. The methods used have, in the main, been based on the known structure of the human ear and consequently, have been based on the use of some form of frequency analyser. However, a frequency analysis alone is clearly inadequate for the recognition even of words, which consist of a definite pattern in time of varying energy content at the different frequencies. Consequently, this approach leads to the attempted recognition of a frequency-time pattern: in effect, of an assemblage of the Gabor information elements[43, 44]. Unfortunately, the pattern of the frequency-time elements varies greatly between speakers; and it may be that in order to make allowance for the filtering effect of the central nervous system on the information from the ear, different approaches will be needed and the emphasis on frequency analysis will have to be abandoned.

While we seem to be a long way still from the recognition of running speech patterns, some progress is being made. However, it is likely that we shall not really be able to set up a machine to recognise human speech as a human does until we have managed to build machines which, given the elements of the information, are then capable of themselves learning to associate the speech patterns with other phenomena. No doubt, when we have devised such learning machines, we shall then learn from them better ways of constructing machines capable of learning speech recognition.

One important thing which must be recognised is the fact that, when hearing speech in a strange accent or in a strange voice or even under new conditions, it is necessary for the human to experience the speech for some time, to become accustomed to the sound, before adequate recognition is possible.

The present position is that, while limited speech recognition is quite possible and has been demonstrated by numbers of people, we are still waiting for a break through in this field. This will in all probability come from someone newly entering the field, who, in ignorance of the immense problems which can be seen by anyone who has spent a lifetime acquainting himself with the difficulties involved, decides to try some very simple approach not at all based on the known knowledge of human speech and hearing.

There has been much work all over the world, and here we shall describe some of the basic work which has been done at Aston, remembering that the basic aim here is to devise speech recognition methods which can be used in conjunction with associating devices such as the Astra machines, so that self-learning rather than mere setting up is the eventual aim.

Vocoder using passive filters

The first practical work on the use of the Vocoder to provide the speech input signals for use with associating machines at Aston was carried out in the Cybernetics Laboratory by A. McMillan and by L. D. L. Soutter[45].

Commercial inductance–capacitance filters were used as the selective elements in the early work. In order to limit costs, commercially available units were used, originally intended for filtering out the signals on voice-frequency telegraphy systems.

Ideally, the band-pass filters used in a Vocoder for experimental use would have a rectangular pass-band. This is not a practicable requirement, however, and instead the filter characteristic was composed of the sum of two, more elementary, filter characteristics for each band.

In order to avoid problems with the phase of the signals from the filters, these were rectified before being combined, as shown in

Figure 15.1. This is quite permissible in the present application, since the output is only required to switch on or off a particular circuit, depending on the energy content in the corresponding waveband of the audio-frequency input.

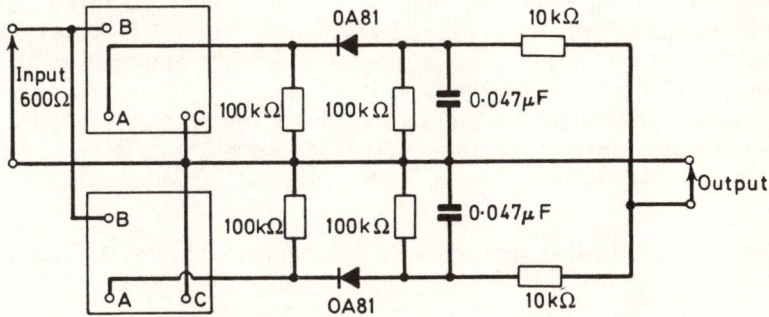

Figure 15.1   Filter-rectifier circuit

The filter arrangement is followed by a potted type of d.c. amplifier, which controls an electronic switch at the output. Thyristors have been used in some forms of output switching device.

Such arrangements have been quite successfully used in initial experiments, but their use has made clear the desirability of considerable size reduction.

Active filters for Vocoder

Various forms of active filter circuit incorporating integrated circuit amplifiers have been investigated for Vocoder use in the Aston

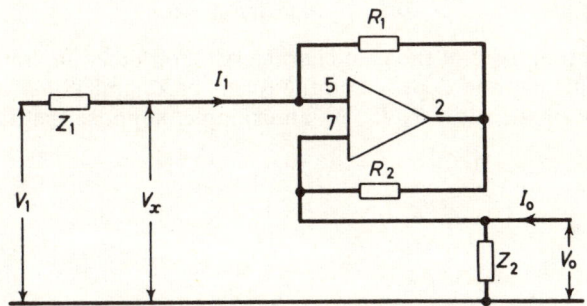

Figure 15.2   Active filter circuit for Vocoder

Cybernetics Laboratory. Only one of these will be described here, the form using negative impedance converters and investigated by P. Harrison[46].

Consider the diagram of *Figure 15.2*. Here an operational amplifier has an output terminal 2, an inverting input terminal 5 and a non-inverting input terminal 7. Now the voltage between the inputs 5 and 7 will always have a small value. Input 5 is connected to the input voltage V_x while input 7 is directly connected to the output voltage V_o. Consequently, $V_x = V_o$.

It can be assumed that no current flows into terminals 5 and 7, so that the whole of the imput current I_1 flows through R_1 and the whole of the output current I_0 flows through R_2. It follows that

$$I_1 R_1 = I_o R_2.$$

Now suppose that an impedance Z_2 is connected across the output voltage V_o. Then

$$I_o = -\frac{V_o}{Z_2}$$

The input impedance is given by

$$Z_i = \frac{V_x}{I_1} = \frac{V_o}{I_1} = -\frac{I_o}{I_1} Z_2 = -\frac{R_1}{R_2} Z_2$$

If now the input to the amplifier comes from a voltage source V_1 via an impedance Z_1, then the output voltage V_o will be given by

$$V_o = V_1 \times \frac{Z_i}{Z_1 + Z_i}$$

so that the effective gain of the circuit is given by

$$\frac{V_o}{V_1} = -\frac{Z_2}{Z_1(R_1/R_2) - Z_2}$$

The actual form of the gain characteristic obtained in any application of this method depends on the forms of Z_1 and Z_2 as well as on the value of the ratio R_1/R_2. As an example, suppose that

$$Z_1 = \frac{1 + j\omega CR}{j\omega C}$$

and

$$Z_2 = \frac{R}{1 + j\omega CR}$$

Then

$$\frac{V_o^*}{V_1} = -\frac{R_1}{R_2} \times \frac{j\omega CR}{1 + j\omega CR(2 - R_1/R_2) + j^2 \omega^2 C^2 R^2}$$

Now the response of a series tuned circuit with the output across the resistor is given by

$$\frac{V_o}{V_1} = \frac{j\omega C_s R_s}{1 + j\omega C_s R_s + j^2 \omega^2 L_s C_s}$$

where the series-tuned components are resistance R_s, capacitance C_s and inductance L_s. Now if the Q factor of the series-tuned circuit is $Q_o = \omega_o L_s/R_s = 1/\omega_o C_s R_s$, where ω_o is the resonant angular frequency, then $C_s R_s = 1/\omega_o Q_o$, so that we can write

$$\frac{V_o}{V_1} = \frac{j\omega/\omega_o Q_o}{1 + j\omega/\omega_o Q_o + j^2 \omega^2/\omega_o}$$

Consequently, the equivalent Q factor of the active filter is given by

$$Q_o = \frac{1}{2 - R_1/R_2}$$

If $R_1 = 2R_2$, then $Q_o = \infty$. Any required value of Q_o can be obtained merely by adjusting the value of the ratio R_1/R_2 between 0 and 2.

Practical robot Vocoder using active filters

A complete Vocoder for use in robots has been designed and constructed in accordance with the foregoing theory by P. Harrison in the Aston Cybernetics Laboratory[46]. The filters incorporate integrated-circuit d.c. amplifiers as shown in *Figure 15.3*. The connections to the amplifiers (type 701c) are as follows:

1   0 V supply
2   output
3   +12 V supply
4   compensation
5   inverting input
6   compensation
7   non-inverting input
8   −12 V supply

262 Robot Hearing and Speech

At the output of each selective amplifier in the Vocoder, a field effect transistor is used, since this form of selective circuit is sensitive to the loading effect at the output. An emitter-follower and a rectifier then give a smoothed output. For some applications a thyristor is used to give a power output.

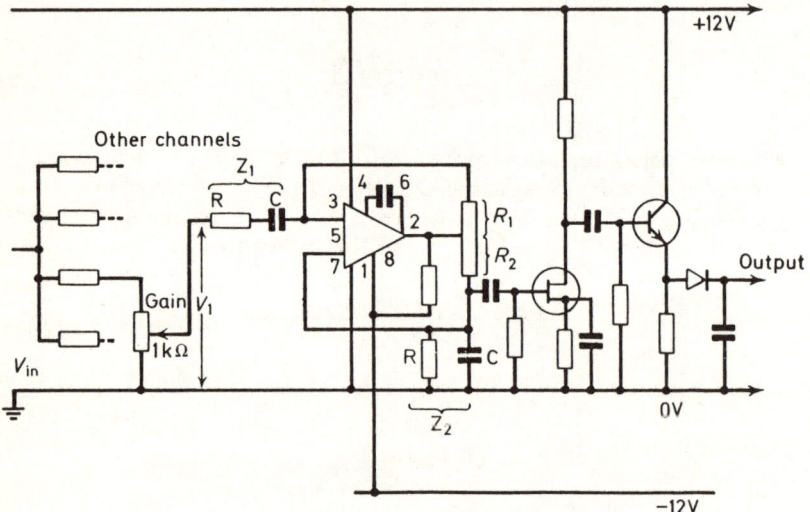

Figure 15.3 Practical robot Vocoder using active filters

With such circuits it is not difficult to obtain quite high value of Q, a typical frequency response being shown in *Figure 15.4*. In some applications the speech waveform has been infinitely clipped before application to the filter circuit. This has the incidental beneficial effect that it prevents amplifier paralysis due to saturation. There is some time-drift in such circuits, but this is not excessive at low values of Q. In some of the work using these filters the differences between the outputs of the various channels have been taken, so as to provide contrast enhancement[17].

Mechanical filters in a Vocoder

It is quite possible to use mechanical filters[1] of the tuning-fork type to carry out the frequency segregation in a Vocoder. In the past, however, this approach has tended to be very expensive and also to give a much sharper frequency response than is necessary. Recently, fibre type optical light guides have been used in order to limit the cost and to provide a simpler over-all solution to the problem. The audio signal is applied to a piezo-electric vibrator which causes the optical light guide fibres to vibrate. Light from a light source is then transmitted through the vibrating fibres on to an optical mask so that a

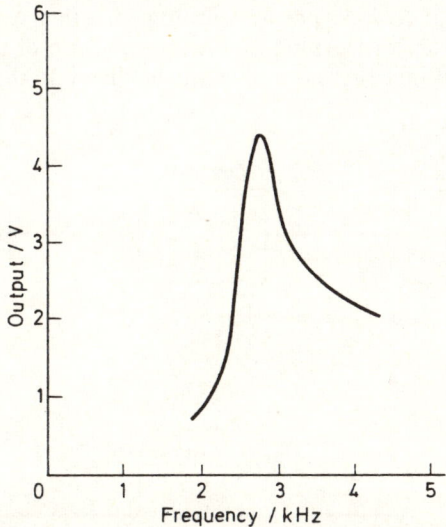

Figure 15.4 Frequency response curve for robot Vocoder

maximum of light is transmitted when certain sounds or words are applied to the input vibrator. An inhibition arrangement is included in order to prevent the operation of the output by noise and interference.

An approach to numeral recognition

Following human experience in, for example, the achievement of powered flight, it is necessary to ask whether the animal hearing system, involving a form of low-sensitivity frequency analysis, might in fact not be the best approach for use in a robot. Unfortunately, if a departure from the animal method is to be adopted, there is a very wide range of possible methods which might be adopted. As with other human achievements, the best answer for use at any given time is only likely to be found by a relatively free search through the possibilities, keeping the objects in mind but avoiding inhibition by any preconceived notions.

Various experimenters have demonstrated the use of the infinite clipping of speech signals which was mentioned earlier. In the

264 Robot Hearing and Speech

Cybernetics Laboratory at the University of Aston D. J. O. Brown has used infinite clipping in the heart of his preliminary form of a recognition system for spoken digits[47]. This system has demonstrated some measure of invariance between speakers—even between male and female voices. It should be stressed, however, that there has been no attempt to incorporate learning into this system. Even so, there is a measure of learning involved in the use of the equipment. It is found that since there is immediate feedback to the speaker, who

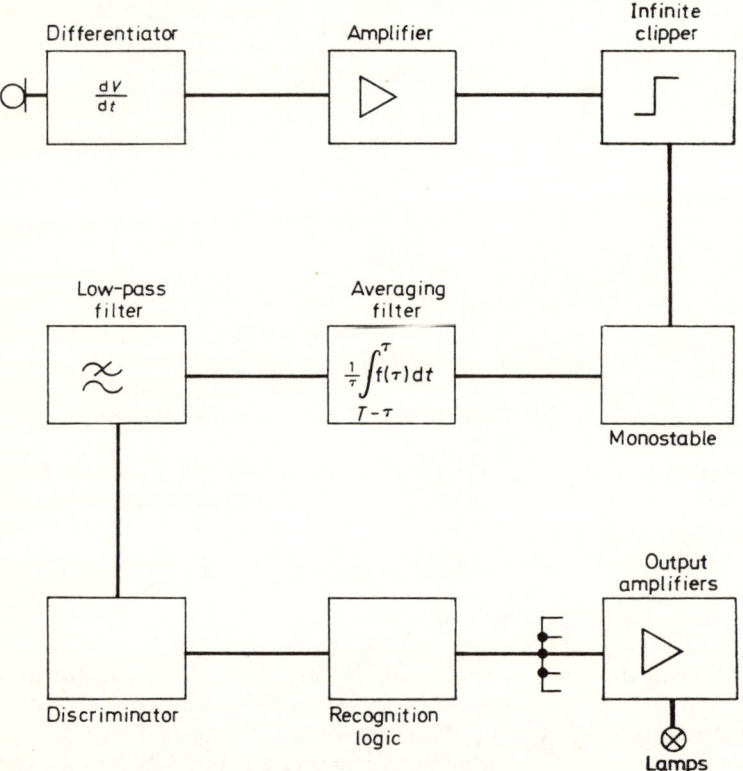

Figure 15.5 Block diagram of speech recogniser

can see whether or not the machine has recognised his spoken word, the speaker modifies his voice slightly in order to make himself more understandable to the machine. The modifications took the form of some emphasis on the plosives and fricatives in the words being spoken. In fact, these are just the forms of modification which a human incorporates into his speech when speaking to another human

Robot Hearing and Speech 265

who is somewhat deaf or who does not understand the language well, or over a poor telephone line. The feedback is thus an important factor in such a speech recognition system.

The system investigated by Brown is shown in block form in *Figure 15.5*. The speech signals from a microphone are taken to a differentiator. This has a response characteristic which rises at the rate of 6 dB/octave from 10 Hz to almost 10 kHz. Above 10 kHz the signal frequencies have little interest and the response is allowed to fall off, again at the rate of 6 dB/octave. After differentiation, the signal is taken to a high-gain amplifier giving over 70 dB gain. It should perhaps be mentioned that the use of integrated-circuit d.c. amplifiers in such an arrangement was found to lead to difficulties, particularly with pick-up, and it proved necessary to construct a special amplifier.

The signal is taken after amplification to a differential comparator, at which it is compared in instantaneous amplitude with a fixed level. The comparator has the effect of acting both as an infinite-clipping device and as a threshold device which removes low-level noise from the system.

The output from the comparator takes the form of a train of pulses having a constant amplitude but varying pulse widths and pulse frequencies. These pulses are applied to an amplifier and then to a monostable multivibrator which gives pulses having a constant width and a constant amplitude each of which occurs whenever the differentiated and infinitely clipped input signal passes through zero in a positive direction. These constant-property pulses are then taken to an averaging filter, which gives the average value over a short fixed interval. This uses integrated circuits in a somewhat complex feedback arrangement which will not be further discussed here. The higher frequencies in the output from the averaging filter are removed by an active low-pass filter having a Butterworth characteristic and a cut-off frequency of 150 Hz.

The output from the filter consists of a slowly varying waveform which is characteristic of the numeral being spoken, but not of the speaker uttering the word. Digital circuits are then used in the final recognition logic in order to light lamps depending on the word spoken into the microphone. The experimental equipment was arranged to recognise the words 'one', 'two', 'six', 'seven' and 'eight', and to light the correct lamp.

While much further work is required on systems such as this, the success of the system in speaker-invariance does indicate the utility of abandoning many preconceived notions in this field. The system can, no doubt, be both simplified and extended if further finance can be obtained for research in this field.

Robot speech[2-6, 19, 23, 34, 35, 38-40, 48, 49]

Although there will be a big advantage in making robots which are directly capable of understanding and acting upon human spoken orders, it is unlikely that robots capable of true speech production will be manufactured for some time.

There is no basic difficulty in the production of some form of human-sounding speech. The obvious way is by the storage of a certain number of prerecorded words, simply recorded by a human and available on a recording medium such as magnetic tape to be called up whenever required. Indeed, this form of speech has been available in telephone systems for a very long time—for example, the prerecorded time announcements (TIM) which are automatically assembled from a number of separately recorded words. Before the doors of the elevator close on the London subway system, a voice tells us, in an accent dating back to the 1930s, to 'stand clear of the doors'.

This form of speech production is very easy to incorporate, and it is therefore likely to be the only form of robot speech for some time. However, it is possible to construct human-sounding speech from elementary functions. This has been done for a long time. Indeed, the Voder, in which human-sounding speech was produced by the manipulation of controls by a human operator, seems to have preceded the Vocoder, in which the reconstituted speech is under the control of the received low-frequency signals produced from the speech at the sending end.

The electrical signals which control the receiving end of the Vocoder could be provided by a robot, which would then be capable of a very flexible form of speech. However, if such a device is to be controlled by a learning mechanism, a long learning period might be required.

At the present time, both forms of speech-producing equipment are made commercially for use with digital computers. However, for the mobile robot it is essential that both the size and the weight of any voice-production unit should be very small to ensure portability.

There appears to have been little work up to the present time on the miniaturisation of either form of audio response unit. The prerecorded type involves mechanical movement, which might be undesirable in a mobile robot. On the other hand, the system is basically simpler than the static Vocoder receiver type. Probably future work will involve the use of specially integrated versions of the Vocoder type to give a small size and weight and to avoid the necessity for moving parts. Early robots will need only a very limited vocabulary of speech, but it is probably better to incorporate the basically more flexible Vocoder system, since this uses the elements of speech sound rather than complete words and, consequently, is capable of further extension.

It should perhaps be mentioned that the robot is, of course, not limited as is the human in the range of output devices for communication which can be operated by the central nervous system. In addition to direct output of modulated radio-frequency signals, there is no reason why a robot should not be fitted with direct visual output by way of a cathode-ray tube, outputs in the infra-red or ultra-violet regions invisible to humans or outputs at ultrasonic sound frequencies. Even permanent forms of output such as printed, punched tape magnetic recordings, etc, can easily be provided if the additional complication and cost and the reduced reliability of the over-all robot system can be justified.

Some recent work on the production of speech as the output from a computer has used phonemes generated by the computer output, coupled together by various audio-frequency 'coupling' signals. The combination of the phonemes with the coupling signals produces a human form of speech. From 100 to 200 different signals are used. In other work the voice source waveforms and the voice channel transfer functions have been simulated by electrical networks[7]. Speech segments have sometimes been assembled to produce speech[8], and speech has also been synthesised successfully by use of simple infinitely clipped zero crossing waves[9]. This method would appear to be promising for computer use.

Detection of self-produced signals

The human makes only a limited use of the detection of signals which he produces himself. For example, one feels one's way in the dark, making an exploratory movement with the hand or foot and detecting the presence of obstacles by their reaction on the probing limb. The writer was once led in pitch blackness by someone who extended both hands in front of him but, unfortunately, omitted to wave them from side to side. The poor chap recovered quite quickly, though the edge of the open door left a nasty vertical bruise on his forehead. The essential feature of such activity should be a widespread exploration which can be narrowed down quickly to focus on any points of particular interest in the environment.

Various animals make use of reflection or reaction of signals produced by the animal itself. For example, the bat produces ultrasonic pulses and perceives its surroundings by the nature of the echo. Similar methods can now be used by the robot. Fishes of some types produce electrical pulses and use these to detect the presence both of obstacles and of other similar fish.

In addition to these natural methods, other methods of detecting the nature of the surroundings by the reflection of self-produced signals are available to the robot. The obvious method is the human one of flood-lighting and of optical detection, and here the flood-lighting can if required use a wavelength of radiation which is not detectable by the human, such as infra-red. Even lower frequencies of radiation, such as radio-frequencies, can be used. Quite small versions of microwave radar devices are now available, and these are suitable for use on the robot[10]. A more complex radar device suitable for robotic applications is the fairly small system used by the police for speed measurement by use of Doppler principles.

Ultasonic radiation through the air or through water has long been used in the form of the ASDIC device for detection of submarines. More recently, such devices have been used in guidance systems for the blind and in industrial applications for the replacement of photo-electric equipment. In the latter form of application, the problem has, in general, been that of price, since a fairly complex transmitter of ultrasound is required in comparison with the simple lamp used for photo-electric equipment. Frequencies of 40-50 kHz are used.

A form of ultrasonic 'eye' having a 10×10 array of elements operating at 455 kHz, together with a 40 cm diameter liquid lens, has been described by Cook[42].

Electrical resistance can be detected and measured by electronic apparatus. While this method has been used industrially, it has, in general, been restricted to use for detection of the levels of conducting liquids such as water and acids.

For the detection of magnetic materials or indeed of any metallic conducting surface, magnetic methods can be used. These vary from the simple detection of the effect of iron by the movement of a magnetically operated switch movement, to the detection of metals such as mines with the mine detector equipment used militarily. Such apparatus has been found to be most useful in industry, where many conducting metals are used, and they will be useful for use with the industrial robot. In robots for military applications such methods will be invaluable, since they can be used to avoid the risk to human life in minesweeping. In even more warlike activities, the radio-frequency proximity detector, in the form of the proximity fuse for use in the nose-caps of shells, has been developed and was used even before 1945. Later forms of guided anti-aircraft weapon have used even more sophisticated forms of detection, both of self-produced signals and of radiation such as infra-red produced by the motive power source of the target.

Robot Hearing and Speech 269

While the magnetic or inductive proximity detector is restricted, in general, to use with metals, detectors using the change of capacitance are not so restricted. Consequently, this form of detector, which involves the continuous measurement of the effective capacitance of a probe and the detection of capacitance changes, has a wide field of application and has been widely used. The disadvantage of the capacitive type of device compared with the inductive proximity detector is the relatively high impedance of the capacitive device. Because of this, such devices are susceptible to interference from stray fields in the environment, and the choice between this form and the inductive form is not always easy.

Another form of self-produced signal which is occasionally used is the radiation from a radioactive source.

It will be noted that, in general, none of the methods discussed can yet give the fineness of detail which can be achieved by optical and television methods.

Inductive and capacitive proximity detectors[11-15]

The inductive proximity detector is suitable for the detection of either magnetic or non-magnetic conducting materials. Many different forms of inductive proximity device have been developed.

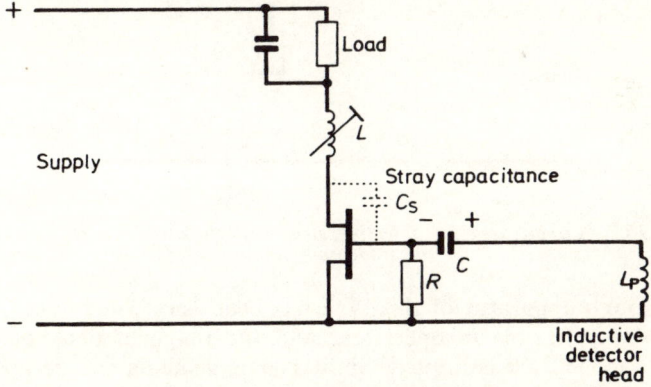

Figure 15.6 Basic circuit for inductive detector

One simple form, which was manufactured in industry by the writer some years ago, used an electronic oscillator, the feedback of which depended on the presence or absence of metallic conducting material from the vicinity of an inductive coil forming part of the oscillating circuit.

The basic circuit shown in *Figure 15.6* forms an oscillator having two tuned circuits without inductive coupling between the two. By circuit transposition as in *Figure 15.6* it is easy to show that in fact the circuit is equivalent to an oscillator having a single tapped resonant circuit.

In the absence of a conductor in the proximity of the detector coil L_p, the circuit oscillates in class 'C', and in these circumstances the average value of the load current is maintained very low by the effect of the bias voltage which builds up across the input capacitor C.

However, if a conductor is brought near to the detector coil L_p, then the effective voltage appearing across the coil falls. If it is reduced sufficiently, then the feedback is no longer sufficient to maintain the oscillation. If, because of this, the oscillation ceases, then the bias voltage across the capacitor C slowly decays as the charge of the capacitor leaks away via proximity inductor L_p and the leak resistor R. The bias voltage falls, and the load current is therefore allowed to increase.

This increase of load current can be used to operate external switching equipment. The presence of conducting material in the vicinity of the inductance coil L_p can in this way be detected and used to operate external apparatus.

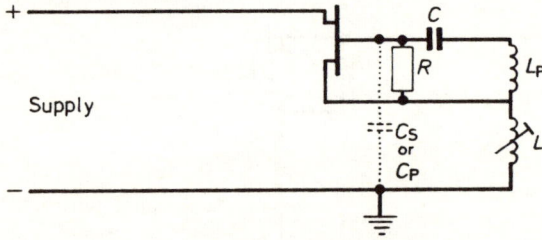

Figure 15.7   Circuit used for detecting stray capacitances between an external electrode and ground

Although apparatus of this type has been very widely used, it is somewhat inflexible in operation, and the range of detector coils which can be used is limited. Similar arrangements can be used to detect changes of capacitance between an external electrode and ground by using a fixed inductance coil L_p but having an external capacitance detector probe C_p connected as shown in *Figure 15.7*. If the value of C_p is increased sufficiently by the presence of any body having a dielectric effect in the vicinity of the detector probe, then the feedback can be reduced sufficiently for the oscillation to cease and for the external switching device to be operated. This capacitive arrangement has the advantage that it can be used to detect a wider range of materials than can the inductive probe arrangement described earlier. However, it has the disadvantage in comparison with the inductive probe that the impedance of the capacitive probe is very high at low frequencies; and, consequently, this form of switch is susceptible to maloperation by interference from low-frequency electrostatic fields.

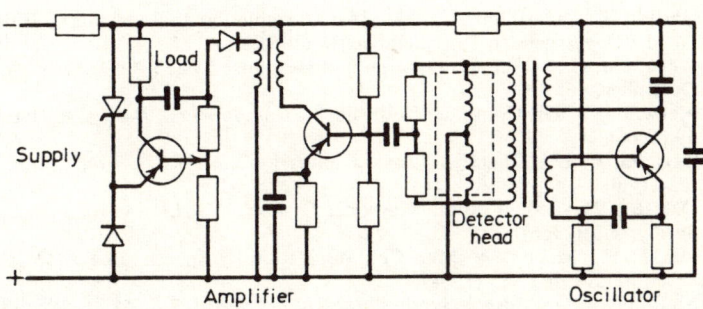

Figure 15.8   Advanced form of proximity detector using bridge technique

A more advanced form of proximity detector makes use of a bridge technique. A typical example developed by D. E. Bench and the writer some years ago is shown in *Figure 15.8*. Here an oscillator operating at high frequency is used to supply a bridge which includes two inductive (or possibly two capacitive) elements. Any unbalance in the bridge caused by the presence of external objects is amplified and then used to operate a switching transistor which can operate external devices. Such circuits are very flexible in operation and they can be used with a very wide variety of detector heads.

With the advent of integrated electronic circuits, the use of arrangements such as those described here becomes very attractive for use in robot touch sense organs, particularly since there is no necessity for actual contact with the object being detected. The reliability of such apparatus is good, though this needs further investigation in the mobile and very changeable environment which will be encountered by the robot, particularly when it is used in industry.

In one form of capacitance detector used industrially the probe capacitance is repetitively charged to a fixed values of voltage and then discharged through a resistor. The average voltage across the resistor depends on the value of the capacitance of the probe. One oscillating inductive detector incorporates an adjustable resistor to vary the operating point and another to vary the backlash or differential between operation and cessation of the output, and temperature compensation is also used.

Radio-frequencies have been used to operate switching devices implanted within the body[16], and to transmit signals from within the body to the outside[18].

REFERENCES

1. Hatfull, T. J. and Jones, G. L., 'An Introduction to Mechanical Filters', *Plessey Component Jl*, **1**, September, 18 (1965).
2. Weitzman, C., 'Voice Recognition and Response Systems', *Datamation*, **15**, December, 165 (1969).
3. Melik, L. F., 'The Computer Talks Back', *Data Process.*, *Detroit*, **8**, October, 58 (1966).
4. Ragland, E. L., 'Digital to Voice Conversion', *A.F.I.P.S. Conf. Proc.*, **20**, 135 (1961).
5. Dale B., 'Never-Fail Audio Response System', *A.F.I.P.S. Proc. Conf.*, **28**, Spring, 277 (1966).
6. Lee, F. F., 'Machine-to-Man Communication by Speech', *A.F.I.P.S. Conf. Proc.*, **32**, Spring, 333 (1968).
7. Mattingly, I. G., 'Speech Synthesis by Rule', *Trans. IEEE*, **AU16**, 198 (1968).
8. Lee, L. H. and Mulvany, R. B., 'Talking Computer Answers Inventory Enquiries', *Electronics*, **36**, August 16, 30 (1963).
9. Sakai, T. and Otani, K., 'Speech Synthesis System Using Zero Crossing Waves', *Electron. & Commun. Jap.*, **52c**, November, 171 (1969).
10. Anon., 'Enter the Hip-Pocket Radar', *Des. Electron.*, **8**, April/May, 5 (1971).
11. Giles, A. F., *Electronic Sensing Devices*, Newnes (1966).
12. Neubert, H. K. P., *Instrument Transducers*, Oxford University Press (1963).
13. Butterworth, A., 'Development and Use of Magnetic Apparatus for Bomb and Mine Location', *JIEE*, **95**, pt 2, December, 645 (1948).
14. Roston, B., 'Development of Locators of Small Metallic Bodies Buried in the Ground', *JIEE*, **95**, pt 2, December, 653 (1948).
15. West, S. S., 'Land Mine Locators' *Electron. Eng.*, **18**, March, 69 (1946).
16. Weller, C., 'Remotely Actuated Solid State Switch', *Electron. Lett.*, **2**, 191 1966).
17. Martin, T. B. and Talavage, J. J., 'Application of Neural Logic to Speech Analysis and Recognition', *Trans. IEEE*, **MIL7**, April/July, 189 (1963).
18. MacKay, R. S., 'Radio Telemetering from within the Human Body', *Trans. IRE*, **ME6**, June, 100 (1959).
19. Van Gieson, W. D. and Chapman, W. D., 'Machine Generated Speech for Use with Computers', *Computers Autumn*, **17**, 31 (1968).
20. Clapper, G. L., 'Automatic Word Recognition', *IEEE Spectrum*, **8**, August, 57 (1971).
21. Clapper, G. L., 'Machine Looks, Listens, Learns', *Electronics*, **40**, October 30, 19, (1967).
22. Clapper, G. L., 'Digital Circuit Techniques for Speech Analysis', *Commun. & Electron*, May, 296 (1963).
23. Flanagan, J. L., *et al.*, 'Synthetic Voices for Computers', *IEEE Spectrum*, **7**, October, 22 (1970).
24. Lindgren, N., 'Machine Recognition of Human Language', *IEEE Spectrum*, **2**, March, 114 (1965).

25. Gilmour, W. D., 'Electronic Speech Recognition', *Wireless Wld*, **75**, February, 76 (1959).
26. Bell, H. A., et al., 'Some Aspects of Speech Recognition for Man–Machine Communications', *IEE Colloquium, London, April, 1968*.
27. Young, John F., *Cybernetics*, Ch. 13, Ilffe (1969).
28. McCarthy, J., et al., 'A Computer with Hands, Eyes and Ears', *Proc. Fall Joint Computer Conf.*, **33**, 329 (1968).
29. Scarr, R. W. A., 'Speech Recognition by Machine—Art or Science?', *Electron. & Power*, **17**, August, 302 (1971).
30. Pierce, J. R., 'Whither Speech Recognition?', *J. Acoust. Soc. Am.*, **46**, 1049 (1969).
31. Reddy, D. R., 'Speech Input Terminals for Computers', *Proc. IEEE Int. Comput. Group Conf., Washington, June, 1970*, 28.
32. Pols, L. C. W., 'Real-Time Recognition of Spoken Words', *Trans. IEEE*, **C20**, September, 972 (1971).
33. Kaiserman, D. B., 'The Awareness and Acceptance of the Audio Response Concept', *Telecommunications*, **5**, October, 16 (1971).
34. Anon., 'Talking Robot-SRI Launch Development Plan', *Electron. Wkly*, January 12, 8 (1972).
35. Poppe, C. W. and Suhr, P. J., 'How Robot Voices Vector Fighter Pilots', *Electronics*, **32**, January 9, 47 (1959).
36. Anon., 'Speech Recognised without Computer', *Electronics*, **45**, June 5, 6E (1972).
37. Hill, D. R., 'Star—A Machine to Recognise Spoken Words', in: *Information Processing 1965, I.F.I.P.*, **2**, May, 357 (1965).
38. Anon., 'Phoneme Phenom', *Electronics*, **43**, June 8, 42 (1970).
39. Flanagan, J. L., 'Synthesis of Speech', *Sci. Am.*, **226**, February, 48 (1972).
40. Denes, P. B. and Pinson, E. N., *The Speech Chain*, Bell (1963).
41. Anon., 'Wheelchair Takes Voice Command', *Electronics*, **45**, September 11, 38 (1972).
42. Cook, R. L., 'Experimental Investigation of Acoustic Imaging Sensors', *Trans. IEEE*, **SU19**, October, 444 (1972).
43. Gabor, D., 'Theory of Communication', *JIEE*, **93**, pt 3, 429 (1946).
44. Young, John F., *Information Theory*, 66, Butterworths (1971).
45. McMillan, A. and Soutter, L. D. L., 'A Vocoder', University of Aston (1968).
46. Harrison, P., 'Speech Recognition', University of Aston (1970).
47. Brown, D. J. O., 'Some Aspects of Automatic Speech Recognition', University of Aston (1970).
48. Levin, H. and Lord, W., 'Portable Data Aquisition Unit for Word Recognition Research', *J. Audio Eng. Soc.*, **20**, October, 656 (1972).
49. Ainsworth, W. A., A Real Time Speech Synthesis System, *Trans. IEEE*, **AU20**, December, 397 (1972).

16

Robot reliability

Finite life of robots[1, 2, 33-35]

The possibility of simply regenerating a complete organism in the same way that some animals regenerate lost limbs appears not to appeal to nature, perhaps because such a process would not give sufficient opportunity for adaptation to long-term changes of environment.

However, when one considers a non-living and non-reproducing robot, is a form of 'immortality' possible here? Is it possible to construct a robot which will last for ever? Obviously, the parts of the robot will wear out in time, but these can be individually replaced.

McNaughton[1] has quoted the treatment of Moore on this question. Suppose that a robot can detect when any part is defective or worn out, and can replace that part from a stock-pile. A finite life of each of the parts of the robot can be assumed, which implies that there is a certain probability of each of the parts lasting some given length of time.

Let the probability of failure of a part of the robot be never less than some constant value k, and let the robot considered be composed of not more than n parts. Then Moore states that the probability that all the parts will fail at once during a time interval is at least the constant $1-(1-k)^n$ and, hence, the probability is zero that in an infinite amount of time all of the parts will never fail at once; that is, that there will never be a total failure of the robot.

In the human being it is an unfortunate fact that as the amount of experience and learning increases, so eventually the facilities such as hearing tend to fail. However, this is not necessarily true of the robot, since replacement of faulty parts is greatly simplified in this case. In addition, with certain types of robot memory storage device it will be possible to transfer the entire memory of an ageing robot to that of a newer robot, and the animal problem of transference of memories from one generation to the next will be greatly reduced. Whether this would result in an undesirably rigid moulding of behaviour patterns remains to be seen.

The investigation of Kletsky[2] into the upper bounds on the life of a self-repairing system indicates that if all components fail at the same given rate whether they are in use or are in reserve, then a self-repair procedure cannot be expected to increase the life of the equipment to more than some three times the mean life of the elements. If, however, the failure rate of stored components in reserve is substantially lower than that of the same components when in use, then the life of a closed and inaccessible system is linearly proportional to the number of spare elements.

It has been suggested that robots left unattended by humans will gradually degenerate and become unable to continue their work because, however much the ability to repair each other's breakdowns is built into them, breakdowns will occur which have not been allowed for by the human designers. It also follows that robots which can build copies of themselves and so are self reproducing will gradually introduce imperfections and faults so that after only three or four generations they are useless.

It is clearly a little early in the history of robotics to attempt to discuss such suggestions. Suffice it to point out here that by the laws of chance it is to be expected that if imperfections are gradually introduced then so will be improvements, purely by chance. This, after all, seems to be the way that animal evolution operates. Using the analogy of evolution, one would expect that there will eventually be a chance of evolutionary improvement in the self-reproducing robot. Nature has made many mistakes after all.

However, we have far too many tasks to carry out in the development of robots of use to mankind to have time for worry about questions which must be, for the present at least, purely hypothetical. At the same time, it is as well that we take care that our robots do obey the Laws of Robotics.

Logical selection

In selecting devices for use in any particular engineering application, it is necessary to take into account a number of factors. In order to ensure that a logical process is followed when comparing various possible devices, it is wise to make use of a definite pre-established procedure. This can take the form of a check list such as the Racet program proposed by Lamb[21, 22].

In this, the word Racer is a mnemonic composed of the initial letters of the headings of a check list. These headings are as follows:

Reliability
Availability
Compatibility
Economy
Reproducibility

These features can be defined in the following way:

Reliability. The percentage of items which continues to operate within a given tolerance range of performance for a given time and under definite operating conditions.

Availability. The percentage of items ordered which are actually delivered on time. Additional factors are the number of different suppliers involved for a particular device type and the length of time between ordering and delivery.

Compatibility. A comparative measure of how well items will operate in the given environment and in conjunction with previously existing equipment.

Economy. The cost of whichever of the types of device under comparison has the lowest over-all cost, is expressed as a percentage of the over-all cost of each type considered. The over-all cost should include initial cost (including purchasing, sales, storage, transport, etc.), test and inspection, installation, maintenance and repair, replacement, etc., where any of these are relevant to the enquiry.

Reproducibility. A comparative measure of the tolerances with which initial characteristics are controlled in production and the resulting degree of interchangeability. Devices having characteristics which are not critical are advantageous, since close control and selection in production are obviated.

Five different grades of quality have been proposed under each heading, with percentage points for commercial and industrial use awarded as follows:

Quality	%
Excellent	20
Very good	16
Good	12
Fair	8
Poor	4

It is possible to prepare quite complex tables of comparison based on the above, but these will not be given here, since, ideally, such tables should be prepared with a specific application in mind.

Schemes of logical selection of components and material for use in the manufacture of complex devices such as robots should preferably take into account other possible characteristics which might be of importance in the particular application being considered.

As an example, when considering equipment to be produced commercially for use in industrial robot applications, the writer has found it useful to extend the Racer mnemonic slightly to TRACERS by adding:

Testability. A comparative measure of the time and cost required for testing equipment in production and in service, and the complexity of the test equipment required.

Simplicity. A comparative measure of the grade and cost of labour required to install and to maintain the equipment.

In effect, these two items have been taken out of the 'economy' category because of their great importance for devices for use in industry, many of which must be handled and maintained by electricians with no special training or equipment.

The effectiveness of a robot

The effectiveness of a robot, or indeed of any other engineering system, can be regarded as dependent on various factors. Some of these are as follows:

1. The reliability. This is usually regarded as being measured by the mean time between failures.
2. The percentage of time that the device is usable, or is ready for use, and the percentage of times that it operates successfully.
3. The quality and availability of the human personnel required to maintain the robot in working order.
4. The average time taken to locate and to repair any failure of the system. This also depends on factor (3).
5. The relative performance under extreme conditions, and when the utmost 'saturation' output is required.
6. The quality and availability of the equipment and facilities required for repair.
7. The spare part requirements per unit time, and the speed with which these spare parts are required to be available. In some cases replacement robots will be available.

It is clear that the relative importance of such factors will depend on the task which the robot normally performs. Consequently, there is no single factor which can be isolated as being of universal importance. It is the task of the cybernetic engineer, at the design stage, to decide on the relative importance of the various factors and to design the robot or the system to meet the requirements as far as is practicable.

Reliability is likely to be an overriding factor in early applications of fully mobile robots. For example, where light alloys are used in the construction, it will be desirable to limit fatigue risks by designing for the ultimate stress to be at least eight times the working stress.

There is likely to be more work in the future on self-repairing robots, particularly with robots for use in space. Even within the solar system, it can take several hours for a radio signal to reach earth from a space vehicle and several hours for a command signal to return. In the event of an emergency, such a delay would be too long for the safety of a space mission.

Work on self-repairing robot devices in general uses the technique of: 'what I say three times is true', in other words two parts of the control system are used to check on the performance of every unit. However, in the event of repeated disagreement the majority rule is accepted, even to the extent of eliminating a disagreeing unit and replacing it with a spare unit.

Typical of such work is the work of Avizienis at the Caltech Jet Propulsion Laboratory, which has been christened Star (Self Testing and Repairing)[5]. In a computer sub-divided into ten separate sub-units, one unit, known as Tarp (Test and Repair Processor), carries out a supervisory function and controls the replacement of suspected defective sub-units, being itself subject to a majority rule.

It is important to remember that robots will have to operate in environments and under conditions which are not suitable for a human being, and in some cases this will make it difficult to obtain a good robot reliability. A typical example of this is the known difficulties with semiconductor control circuits in environments of high atomic radiation. Other problems are those of an electrically noisy environment[3, 19, 20, 24, 26, 32], mechanical vibrations[27], and magnetic interference[28].

Mean time between failures[25]

As the complexity of manufactured equipment, both mechanical and electrical, has increased, so the subject of reliability has become more and more important. At one time it was not unusual to take the naive viewpoint of ignoring the possibility of failure. However, when the failure of complex equipment can cause the loss of a battle, or the expenditure of many thousands of pounds in industrial downtime cost, it is no longer possible to ignore the fact that breakdowns can and do occur. Nothing lasts for ever.

The aspect of reliability that is most important depends on the particular nature of the application of the device being considered. For example, in process control equipment the relative amount of downtime is important. In military equipment the percentage of operational missions which are successfully completed is all-important. In other equipment, and perhaps robot equipment should be included in this category in many cases, the important feature is the mean time between failures, sometimes known as the 'reliability index'.

However, in order to determine the likely mean time between failures, it is necessary to carry out life tests on actual equipment under normal operating conditions. This can take a very long time, and the incorporation of any changes which are found to be necessary in production can be very expensive.

Consequently, it is of interest to consider the actual pattern of failure during life which is encountered, in case this can give additional information on the nature of the reliability to be expected. The reciprocal of the mean time between failures is known as the failure rate.

Failure rate curves

The rate at which any equipment fails in service is not constant during the life of the equipment. Instead there is an initial period when the failure rate is large. This is often known as the 'burn-in' period. There is then a long period when the rate of failure is very low. Eventually parts begin to wear out and the failure rate starts to increase again. If the rate of human failure is plotted against time, a 'bathtub' form of curve is obtained. Such curves can be plotted for the human body as shown in *Figure 16.1*. There is a high infant mortality rate, followed by a reasonably static rate, and then the life of the human is terminated in some way[6]. It has been suggested that such curves illustrate well the fact that heart maintenance is in need of improvement.

Now it is an interesting fact that the failure rate of engineered equipment appears to follow just such a curve. The initial 'burn-in' period is, fortunately, under the direct control of the manufacturer of the equipment, and it is therefore possible to eliminate much of this initial failure by running the equipment for some time at the factory before it is consigned to the customer.

While it is possible to eliminate initial failure in this way, it is very expensive, since it requires a great deal of time and space which must

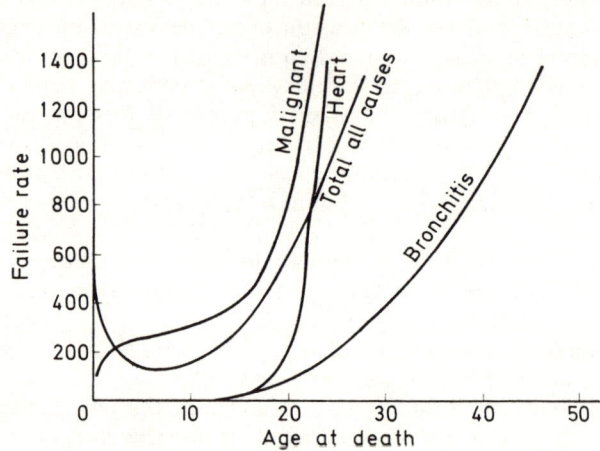

Figure 16.1   Catastrophic failure rates for human males, 1959

be allocated by the manufacturer and which must therefore be paid for by the customer.

Failures during life are, in general, governed by two forms of probability distribution, Poisson and Gaussian.

Poisson distribution of failures

If a robot, or any other engineering system, comprises a total of n separate parts, and if each of these parts can be considered as having all equal reliabilities (or probabilities of operating within the specified performance limits for a specified time under specified operating conditions) each equal to R, then the over-all reliability of the complete system can be said to be equal to R^n.

Now normally the information available to a designer is in the form of mean time between failures or perhaps in terms of the reciprocal, the failure rate. In order to convert such information into the form of reliability information, the Poisson probability distribution law is used. This law gives the probability of a certain number of events occurring in terms of the average number of events which have occurred in the past, and it can be used to predict the probability that a given number of components will fail during a given interval.

Let the probability of failure of any one of the n separate parts of the system be P. Then the number of failures to be expected is equal to nP during the course of the time interval being considered. If a unit time is considered, then the faults per unit time is equal to nP, on average.

Now the Poisson probability series is defined as

$$P_x = \frac{(nP)^x}{x!} \exp(-nP)$$

Expanded, this becomes

$$\sum P_x = \exp(-nP) + nP \exp(-nP)$$
$$+ \frac{n^2 P^2}{2!} \exp(-nP) + \frac{n^3 P^3}{3!} \exp(-nP)$$
$$+ \frac{n^m P^m}{m!} \exp(-nP) + \ldots$$

Now this series contains, term by term, the following:

$P_0 = \exp(-nP) =$ probability of 0 failures per unit time
$P_1 = nP \exp(-nP) =$ probability of 1 failure per unit time

$$P_2 = \frac{n^2 P^2}{2!} \exp(-nP) = \text{probability of 2 failures per unit time}$$

....

$$P_m = \frac{n^m P^m}{m!} \exp(-nP) = \text{probability of } m \text{ failures per unit time}$$

Note that this series must terminate at $m = n$, since there is only a total of n components. Also, since probabilities are being summed, the sum must be equal to unity, $\sum P_x = 1$.

On this basis, the probability that the system will not fail during the given unit time, i.e. that there will be no failures in the unit of time considered, is equal to the first term, $\exp(-nP)$. This term is sometimes taken as a measure of the reliability of the system: $R = \exp(-nP)$.

In the above, unit time has been considered. If the time considered is instead equal to T, then the over-all reliability is equal to $R = \exp(-nPT)$. Here the failure rate is equal to P, the probability of failure of each component per unit time.

282 Robot Reliability

The mean time between failures is equal to the reciprocal of the number of failures per unit time, or

mean time between failures = $M = 1/nP$

Consequently, the reliability can be expressed in terms of the mean time between failures as $R = \exp(-T/M)$. The value of R is plotted against T/M in *Figure 16.2*.

It should be noted that, if the exponential is expanded,

$$R = \exp(-T/M) = 1 - T/M + \frac{T^2/M^2}{2} - \frac{T^3/M^3}{3} + \dots$$

Consequently, if the value of the time interval T is much less than the value of the mean time between failures M, say less than 0.1 M, then the reliability can be taken as approximately $R = 1 - T/M$. Note

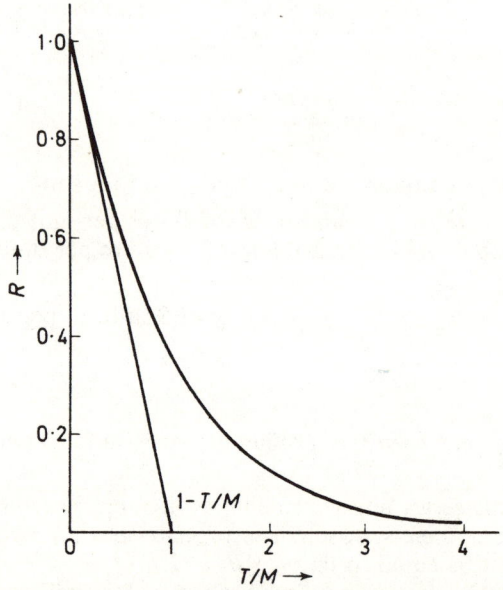

Figure 16.2 Relationship of mean time between failures, R v. T/M

that the value of T/M gives the probability of a failure occurring during the time interval T.

In the above it has been inherently assumed that all components are identical. In the general case this is not so, and it is then necessary to express the mean time between failures as

$$M = 1/\sum n_a P_a$$

As an example of failure rates and mean time before failures, consider a device having 2500 parts, each having a stated failure rate of 0.01% per 1000 hours. Then the failure rate of the system is

$$2500 \times \frac{0.01}{100\,000} = 2.5 \times 10^{-4} \text{ per hour}$$

On this basis, the mean time between failures is the reciprocal:
$$M = 1/2.5 \times 10^{-4} = 4000 \text{ h}$$
This corresponds to about one failure every 6 months. In 1 year, or approximately 8000 h, the reliability is given by

$$R = \exp(-T/M) = \exp(-8000/4000)$$
$$= 0.135$$

In this case the probability that the equipment will operate successfully for 1 year is expressed as 13.5%.

Unequal failure rates

In most cases the parts used in a system will not all have equal rates of failure. The effect of unequal failure rates can best be illustrated by an example.

Consider a system comprising five parts having different values of mean time between failures (MTBF) as given below:

Part	A	B	C	D	E
MTBF (h)	1000	1200	1500	1500	2000
Failure rate (%/1000 h)	0.1	0.0833	0.066	0.066	0.5

The over-all MTBF here is obtained from

$$M = \frac{1}{0.001 + 0.000833 + 0.000666 + 0.000666 + 0.000500}$$

$$= \frac{1}{0.00366}$$

$$= 273 \text{ h}$$

Where there is a definite maximum repair time or downtime t required, and this value of maximum time t is known, then sometimes a figure of percentage availability A is defined as

$$A = \frac{M-t}{M} \times 100\%$$

284 Robot Reliability

In the example quoted, if the maximum repair time t is 12 h, then the percentage availability is given by

$$A = \frac{273-12}{273} \times 100 = 96\% \text{ availability}$$

The importance of such figures depends on the application. For domestic use, a low figure of availability might not be too serious. However, for a high production line in a factory, it would be disastrous because of the high cost of downtime.

Gaussian distribution of failures

The Poisson distribution of failures discussed above can be applied in all cases where it is possible to assume that the average rate of failure is constant, and also where only a small proportion of the total components of the device actually do fail. However, in some equipment it must be assumed that there is some force causing failure of components in a time which is dependent on the ability of the component to resist the force causing the failure, and also that this ability varies in a statistically normal fashion.

The Gaussian probability function is expressed mathematically as

$$\text{probability density} = \frac{1}{s\sqrt{2\pi}} \exp\left[-\tfrac{1}{2}\left(\frac{t-t_m}{s}\right)^2\right]$$

where s is the standard deviation of failure distribution; t is the duration of service; and t_m is the mean life, or mean of failure distribution.

If the value of the probability density of failures is plotted against time t, a 'normal' distribution curve is obtained as shown in *Figure 16.3*. This curve is applicable in the cases discussed above.

Since the Gaussian distribution is widely applied, tabulated values are available in statistical tables. In these a 'standard normal curve' is assumed, in which $s = 1$ and $t_m = 0$, so that

$$\text{Probability density} = \frac{1}{\sqrt{2\pi}} \exp(-\tfrac{1}{2}t^2)$$

Then the area under the curve represents probability and the total area and total probability is equal to unity. Statistical tests are used to determine whether the Poisson or the Gaussian distribution should be applied in a given case.

Robot Reliability 285

Over-all failure pattern[7,8]

If the over-all failure pattern of manufactured equipment is examined, it will be seen to contain some features which can be described by the Poisson distribution and others which can be described by the Gaussian distribution. It is an interesting fact that this is true not only of manufactured equipment such as robots but also of living creatures such as man. Indeed, because of the greater numbers available to a given fixed basic design, there is greater evidence for the form of failure pattern in the case of the human being than there is for any manufactured equipment. The required information about failure in the human being is readily available to medical men in statistical form, and typical results were plotted earlier in *Figure 16.1*.

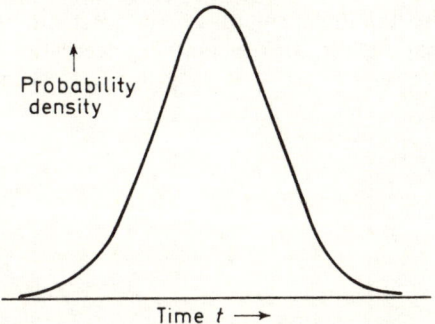

Figure 16.3 Probable density of failures plotted against time

The Poisson distribution can be applied, in general, to random failures which occur very early in the life of equipment. Thorough production testing can eliminate many of these faults. The Poisson distribution also applies to random failures which can occur at any during the life of well-designed equipment. Such random failure can be minimised but never completely eliminated.

The Gaussian form of distribution also appears to apply in two cases of failure of equipment. The first is due to faulty design of the equipment—for example, the operation of components at excessive temperatures due to overrating or to the provision of inadequate ventilation. Such faults inevitably occur because of human design error, and they should be eliminated early in the production life of equipment, provided that the equipment is manufactured in sufficiently large quantities. The second way in which the Gaussian form of failure distribution appears is in the case of 'wear-out' failures. These should not normally appear until the end of the life of equipment is approached, when the number of failures usually shows a

286 Robot Reliability

dramatic increase. It is possible to reduce this effect by careful maintenance, but it has been suggested by Moore[1] that it can be proved theoretically that, even so, an infinite life is quite impossible.

The four factors which have been discussed all contribute to the over-all pattern of failures which will be encountered during the life of any robot. The resulting pattern is often expressed in the form of a 'boat' or 'bathtub' curve of number of failures against time as shown in *Figure 16.1*. It is seen that, as in the case of the human being, there is an initial high rate of failure, followed by a plateau during the working life of the equipment. At the end of the life the number of failures begins to increase drastically.

With well-engineered equipment the height of the central plateau region is small. As production proceeds, because of the detection and elimination of sources of failure, the initial downward slope becomes steeper and the commencement of the plateau becomes earlier, while the actual height of the plateau decreases. This can only

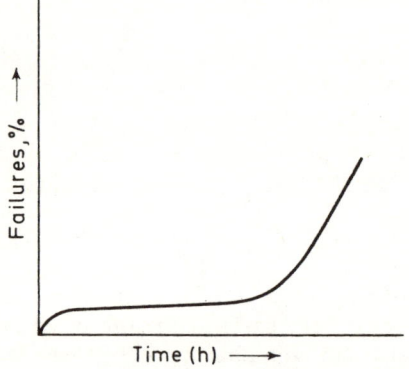

Figure 16.4 Curve showing number of failures against time

happen if there are very long production runs of standardised equipment, however, and it is important therefore that robot equipment should be as standardised and as general-purpose as possible, and that the designer should be 'forward-thinking'.

The failure pattern is normally seen plotted as the bathtub curve of failure rate against time. However, a curve showing number of failures against time can be more informative. Such a curve is shown in *Figure 16.4*. Here the failure rate is equal to the slope of the curve, and the total number of failures is the final point on the curve and is a fixed quantity in any given case. The object of good design is, then, to keep the integral of the curve as low as possible by keeping the whole curve low for as long as possible. However, assuming that some burn-in failures are unavoidable, the object then should be to keep the central region as flat as possible.

It should be noted that the value of MTBF obtained from practical testing can depend very much on the period adopted between tests. For example, it is quite possible to obtain apparent figures for MTBF as a result of monthly testing which are twice as long as those obtained if tests are carried weekly. Care must be taken that results for comparison are always obtained under identical conditions.

Redundant parts[9-15, 23, 29, 31]

One way of improving the over-all reliability of a system or of a robot is by the provision of two or more parallel units to perform every function. This method is used to some extent in the human and animal body, where there are two hands, two eyes, etc. In the case of a robot the obvious problem of economics arises, since the reduction of downtime must be balanced against the increased initial cost, and perhaps also against the need for somewhat complex testing to separate out the faulty redundant parts for repair. An additional factor in the case of the mobile robot is that the weight of the additional redundant units must be considered. This is particularly important in the case of an airborne or a space robot.

Where there is a number n of units operating in parallel, having probabilities of failure of

$$f_1, f_2, f_3, \ldots f_n,$$

then the over-all failure probability F is given by the product:

$$F = f_1 \times f_2 \times \times f_3 \ldots \times f_n$$

This statement should be compared with the reliability probability R of successful operation of a system as discussed earlier. Any unit or any system must either fail or not fail, so $F + R = 1$. Similarly, for any part or component c, $f_c + r_c = 1$. Now $R = \exp(-T/M)$, and, similarly, for any part or component c, $r_c = \exp(-T/m_c)$, where m_c is the mean time between failures of that part c.

If all parts have identical failure probabilities equal to f_c, then

$$F = f_c^n = (1 - r_c)^n$$

Hence,

$$R = 1 - F$$
$$= 1 - (1 - r_c)^n$$
$$= 1 - \left(1 - \exp\left(-\frac{T}{m_c}\right)\right)^n$$

288 Robot Reliability

As an example of the application of this, if units are triplicated, i.e. if three identical parallel units are used and each has an MTBF of 100 h, then the reliability for 1 day of 12 h (i.e. the chance of completing 12 h work without failure) is equal to

$$R = 1 - \left(1 - \exp\left(-\frac{12}{100}\right)\right)^3$$
$$= 1 - (1 - 0.8869)^3$$
$$= 0.998 \text{ or } 99.8\%.$$

This figure compares with a reliability of 98.7% if units are duplicated and 88.7% for a single unit. There is unlikely to be any advantage in the use of more than three parallel parts, as can be seen by plotting the reliability against the number of parallel parts for any given case. However, in some cases triplication would be very desirable, an example being automatic landing equipment for aircraft. The above treatment can be extended for use where faults are found and repaired as they occur.

Walter has pointed out[16] that it is a surprising fact that the over-all failure rate of the human in the form of nervous breakdown is only of the order of 10%, and he suggests that the simple fact of high redundancy in the nervous pathways is insufficient to explain the great reliability. Walter suggests that redundancy is required but that, in addition, an independence of the action of each channel is also necessary. For example, it is important that the power supply to the nervous system should take the form of a large number of quite independent smaller supplies, so that failure of any one of them cannot cause a complete failure. It will be necessary to overcome by such means the fact that the reliability falls rapidly as the number of elements increases. At the present time the only known method of achieving this is by the use of a control system for each part which is quite separate from that for each other part. The necessary cross-coupling between sub-sections can be as far as possible by way of the environment.

In the robot such a sub-division can help to lead to an easy repair and replacement procedure, and so to a reduction of downtime. However, it is likely to incorporate a penalty in the form of an increased size and weight, since at the very least the required sub-units must have some form of individual carrier or container, as well as some form of individual plug-and-socket form of connection with the main body and with the other sub-units. The connections can themselves introduce unreliability, and it is therefore necessary to make a very difficult decision on the extent of sub-division to be adopted[17].

Repair time

The total time taken for repair of a defective robot or other equipment can be regarded as the sum of three component times. These are:

1. The diagnosis time, or the time taken to detect the occurrence of the fault.
2. The time taken to remove the fault from the system.
3. The time taken to repair the faulty item or to replace the faulty module.

In order to obtain the optimum utilisation from the robot, it is necessary to take steps to ensure that all three components are minimised. This can be done by making the procedures as automatic as possible. All three components should be separately considered at the design stage and at the redesign stage to ensure that the best arrangements are made.

It is desirable that any robot should be so arranged that sudden and drastic failures are avoided wherever possible. As Jordan has pointed out[18], one aspect of human flexibility is the capacity for graceful degradation. In other words, although to err is human, spectacular errors such as those encountered with faulty computer systems are fortunately rare, and with ageing or with illness the onset of faulty operation is only gradual. Such a feature is desirable in a robot.

REFERENCES

1. McNaughton, R., 'The Theory of Automata, a Survey', *Adv. Comput.*, **2**, 379 (1961).
2. Kletsky, E. J., 'Upper Bounds on Mean Life of Self-Repairing Systems', *Trans. IRE*, **RQC11**, October, 43 (1962).
3. Dennehy, W. J., *et al.*, 'Digital Logic for Radiation Environment', *RCA Rev.*, **30**, December, 668 (1969).
4. Rexrode, L. O., 'Noise Proof a Digital Electronic Control System', *Control Eng.*, **15**, November, 66 (1968).
5. Anon., 'A Star is Born', *Time*, December 7, 34 (1970).
6. Hardy, D. R., 'The Quantitative Approach to Reliability', *Plessey Component Technol.*, **1**, December, 26 (1964).
7. Siemaszko, Z. S., 'Reliability in Action', *Electron. & Power*, **16**, June, 223 (1970).
8. Mair, B. A., 'Reliability and Accelerated Life Testing', *Plessey Component Technol.*, **2**, June, 5 (1966).
9. Rankin, K. F., 'System Engineering for Reliability and Ease of Maintenance', *JIERE*, **35**, 67 (1968).
10. Van Staalduinen, J., 'Improving Reliability of Amplifiers by Redundancy', *Electron. Eng.*, **42**, June, 17 (1969).

11. Darnell, P. S., 'Electronic System Reliability—An American Viewpoint', *Proc. IEE*, **111**, February, 284 (1964).
12. Jenson, P. A., 'Quadded Nor Logic', *Trans. IRE*, **R12**, September, 22 (1963).
13. Wilcox, R. H. and Mann, W. C., *Redundancy Techniques for Computing Systems*, Spartan (1962).
14. Inskip, F. A., 'Redundancy in Digital Systems', *Electron. Eng.*, **39**, April, 244 (1967).
15. Higgins, J. C., 'Problems in the Specification and Assessment of Electronic Equipment Reliability', *Proc. IEE*, **113**, September, 1413 (1966).
16. Walter, W. G., 'The Past and Future of Cybernetics in Human Development', in: Rose, J. (ed.), *Progress of Cybernetics*, Vol. 2, 55, Gordon and Breach (1970).
17. Flatau, G., 'Reliable Contacts and Connections in Telecommunications Plant', *Trans. Instn Engrs Aust.*, **EE3**, March, 59 (1957).
18. Jordan, N., 'Allocation of Functions between Man and Machines', *J. App. Psychol*, **47**, 161 (1963).
19. Armstrong, D. R., 'TTL Interfacing with GRL 111 and GRL 101', *Mullard Tech. Comm.*, **11**, July, 130 (1970).
20. Simmons, B. D., 'Designing Noise Immunity into Electronic Circuits and Systems', *Electron. Equipt News*, April, 74 (1966).
21. Lamb, J. J., 'Evaluating Electronic Components for Reliability Plus', *Electl Mfg*, December, 111, (1955).
22. Lamb, J. J., 'Racer, a Proposed Rating System for Electronic Components and Devices', *Trans. IRE*, **RQC6**, February, 1 (1956).
23. Chestnut, H., *Systems Engineering Methods*, Wiley (1967).
24. Hormuth, G. A., 'Minimising Electrical Interference', *Mach. Des.*, **43**, April 29, 48; and June 10, 132 (1971).
25. Ask, F., 'Reliability Definitions for Electronic Equipment', *Electron. Eng.*, **42**, June, 13 (1969).
26. Bond, N., 'RFI—One of the Undesirables', *Electron. Eng.*, **43**, August, 32 (1971).
27. Mitchell, W. S. E., 'Industrial Vibrations, Their Effect on Electronic Equipment', *Des. Electron.*, October, 22 (1966).
28. Anon., 'Low Frequency Magnetic Shielding: Shield Fabrication', *Des. Electron.*, February, 14 (1969).
29. Short, R. A., 'The Attainment of Reliable Digital Systems through the use of Redundancy', *IEEE Comput. Group News*, **2**, March, 2 (1968).
30. Wilcox, R. H. and Mann, W. C. (eds.), *Redundancy Techniques for Computing Systems*, Spartan (1962).
31. Avizienis, A., 'Fault-Tolerant Computing, An Overview', *Computer*, **4**, January, 5 (1971).
32. Butenbach, R. W., 'Ground Circuits', *Instrum Control Syst.*, **42**, November, 135 (1969).
33. Briley, B. E., 'A Self-Healing Control', *B.S.T.J.*, **47**, 2367 (1968).
34. Avizienis, A., 'Design of Fault-Tolerant Computers, *Fall J.C.C.*, **31**, November, 733 (1967).
35. Taylor, M. G., Reliable Information Storage in Memories Designed from unreliable Components', *B.S.T.J.*, **47**, 2299 (1968).

17

The future of the robot

As can be seen from the work described in this book, there is little doubt that the robot will soon be with us. Elementary robots are already making their appearance in industry, but these are rather special-purpose devices. Soon, however, probably from Japan, we shall see the introduction of the general-purpose mobile robot. The Japanese Government is investing some 1 000 000 dollars in this field over the next few years, following the successful Hitachi work described earlier[1]. No doubt this is partly due to the full-employment conditions in Japan, though by the nature of the Japanese economy we can expect to see the resulting robot devices exported to the West, for use in industry (and perhaps in domestic applications, where the potential market is immense.) The U.S. Government has granted 1 000 000 dollars to M.I.T. for investigation of Naval robots[2]. The British Science Research Council has granted similar sums for work in 'machine intelligence'[3].

As soon as it is realised that the robot is very suitable for housework, production will soar and prices will plummet. This will be the dangerous time, since the introduction of many different models will be economically suicidal. Let us hope that the international robot will appear, following the lesson learned by the automobile industry and by the electronics industry. The possible future of the robot has been analysed by Huber, using the Delphi questionnaire method[4].

The economic consequences and the social consequences of the introduction into industry of the general-purpose robot will be extreme, and will make present redundancy seem like an insignificant irritation. The speed with which we can expect the widespread introduction of the robot to proceed will be the real problem. Human beings will have little time to adapt slowly to the robotic revolution.

292 The Future of the Robot

What will our robots look like? Prediction is dangerous, but the writer would here make a plea for extreme care. While the outer appearance of the robot will change with time and fashion, the domestic robot is going to be our ever-present companion, and the best possible aesthetic thought should be brought to bear upon its appearance. And the sound and the smell of the robot should not be overlooked at the design stage either.

When our general-purpose robots are introduced, we shall certainly find new and unexpected uses for them. This is just the process of human ingenuity. We have already encountered the process even in the development of our early ASTRA Mk 3 robot brain[5, 6], and the unexpected applications are certain to multiply as the human becomes familiar with his robots.

The robot is a machine. Let no one doubt that, though almost certainly some humans will adopt them as domestic pets, since this is the nature of some humans. The very standardisation which will be necessary in the low-cost general-purpose robot will help us to realise even more forcefully the great variety of human behaviour and appearance. Let us hope that it helps us to greater tolerance of one another.

REFERENCES

1. Anon., 'Japanese Robot Assembles Components from Drawings', *New Scient.*, **54**, September 9, 575 (1971).
2. Anon., 'MIT Lab to Study Machine Intelligence', *Control Eng.*, **19**, June, 43 (1972).
3. Anon., 'SRC Computing', *Sci. Rev.*, June, (1972).
4. Huber, R. P. O., Analysis of the Future of Robots and Artificial Intelligence', *Proc. 1st Conf. Industrial Robot Technology, Nottingham, March, 1973*, 239.
5. Young, John F., *Cybernetic Engineering*, Butterworths (1973).
6. Young, John F., 'Progress with the ASTRA Principle', *IEE/IERE Colloquium On Computer Structures for Artificial Intelligence, London, May, 1973*, 3

Index

Acceleration feedback, 39
Accelerometers, 39
Accommodation, phenomenon of, 17–18
Actuators,
  electrical 59–62
  electro-hydraulic, 121, 140
  exoskeleton, 147
  general requirements, 57
  hydraulic, 81–83, 153
  linear, 153
  pneumatic, 59, 74–81, 84
    limb control, 78
    prosthetic devices, 77
  power source, 58
  rotary, 153
  rotary hydraulic, 147
  safety requirements, 57–58
  solenoid, 63–65
  telescopic, 153
Adaline, 201
Aesthetic Law, 4
Aircraft, 149
  radio-controlled, 116
  target, 116
Alignment problems, 163–164
Alston, B. J., 103
Aluminaut vehicle, 182
Alvin, 182
AND gate, 209
Android, definition, 2
Angular acceleration, 150
Angular acceleration sensor, 148
Angular velocity sensor, 148
Angyan, A. J., 102
Ankle joints, 154
Anthropomorphous machines, 157
Arcton-13, 83

Argonne National Laboratories, 117
Arms,
  artificial, 82, 146
  computer-controlled, 126
  mechanical, 129
  performance requirements, 45
  prosthetic, 175
  requirements, 46
  robot, 127, 139, 146, 166, 175, 224
  telescoping form, 147
ASDIC device, 268
Asimov, I. 2
Assembly machines, automatic, 176
Aston Cybernetics Laboratory, 102, 202, 212, 258, 264
'Astor' mobile machine, 102–104
Astra machines, 12, 206
ASTRA Mk. 3 robot brain, 200, 292
Audible illusions, 251
'Automatic Apprentice', 15
Automatic production lines, 6
Automation, 163
Autoplace, 176
Availability, 276
  percentage, 283–284
Avizienis, 278

Backlash, 52, 80, 81
Backlash circuit, 53
Balance sensors, 148, 149
Ballinger, H. A., 124, 126
Banks, L. K., 100
Barlow Disc motor, 61
Bathyscaph, 181
Batteries, 59, 61, 83–84, 146
  charging, 87

294 Index

Batteries *continued*
  current-voltage characteristic of charging devices, 88
  drain prevention, 72
  economy of operation, 72–73
  general requirements, 87
  maintenance, 87
  mechanically rechargeable, 85
  rapid recharging, 88
  recent developments, 84–86
Battery electric vehicles, 107
Bayes net, 201
Beetle mobile manipulator, 119
Behaviour patterns, 275
Belcher, J. U., 177
Bench, D. E., 271
Bendix Company, 110
Bernard, Claude, 13
Binary output, 214
Binocular vision, 192–194, 237
Blakemore, C., 196
Bolie, V. W., 244
Boni, G., 48
Bottomley, A. H., 47
Boucherot circuit, 38, 99, 100
Boucherot, P. M. J., 100
Braille alphabet, 24
Brain, robot, 127
Brakes, 61, 73–74
Breakdowns, 275, 279
Broadcasting service, 15
Brooke, D. W. 114
Brown, D. J. O., 264, 265
'Burn-in' period, 279
Butane gas, 78

Capsub, 181–182
Carbon dioxide, 75, 84
Carsbury, 201
Cascade addition of equalisation network, 92
Cavanagh, P. R., 152
Cavitation, 83
Cells; *see* Batteries
Character recognition, 200–211
  minimal, 212–215
  optical, 202
  visual, 200–201
Chart recorders, two motion, 228
Chevrolet Vega car, 86
Chitty, A., 77
Clarke, J., 100

Climbing robot, 149, 154, 155
Closed-loop response, 95, 96
Closed-loop stability, 93
Clutches, 61, 73–74
Clynes, M., 192
Coal mining, 16
'Cocktail party effect', 250
Compatability, 276
Computer control, 126–130, 225
  artificial arms, 126
  storekeeper, 150–151
  urban transit vehicle, 111–112
Computer program, Mk-17, 205
Computer scanning system, edge detection in, 224
Computers, 4, 6, 8, 17, 118, 200, 202, 204, 206, 208, 212, 215, 245, 266, 267, 278
Connole, Anthony B., 13
Conslarm, 176
Contini, R., 152
Contrast enhancement, 26, 216
Control console, 145
Conveyors, 163
Cook, 268
Cool, J. C., 81
Cooling, 68–70
Cooper, F. S., 247
Cooper, G. F., 196
Copilia, 229
Copying robots, 8
Cornell Aeronautical Laboratories, 147
Counting retina, 219–221
Crawling machines, 126
Curv device, 183
Cutler, C. C., 107
Cybernetics, use of term, 1

Davies, B. L., 61, 83
Davies, R. J., 238
Dead zone, 73
Degradation, 289
Delay device, 53
Differential comparator, 265
Direction-sensing devices, 194
Discrimination, 204, 206
Domestic robot, 3, 4, 5, 63, 107, 180, 181, 291, 292
Donald, A., 77, 78, 88
Doppler applications, 115, 239
Drilling rigs, 183

Index 295

Drives,
  electrical, noise of, 62–63
  worm-gear, 72
Dudley, H. 243, 246
'Dunmore' sensor, 28

Ear,
  human, 242, 243, 244, 251
  inner, semicircular canal system of, 148, 153
  middle, 243
  robot, 253
  see also Hearing
Earle, B., 100
Economics, 9–12
  of Unimate, 178
  of Versatran, 177–178
Economy, 276
Edge-detecting retina, 215, 221–223
Edge detection, 33, 127, 215, 220
  in computer scanning system, 224
Edge enhancement, 218
Effectiveness of robot, 277
Elbow joint, torque/speed curve, 42
Elbow-torque requirement, 46
Electric motors, 60, 61, 62, 66
  double-stator squirrel-cage, 100
  field control, 98
  for mechanical actuation of robot limbs, 59
  series, electronic control, 98
Electrical actuation, 59–62
Electrical drives, noise of, 62–63
Electro-hydraulic actuators, 121, 140
Electrolytic devices, 149
Electromagnetic sensor, 149
Energy storage, 58, 77–78, 81, 83
Environment,
  hazardous, 139, 278
  industrial, 180
  special, 15–17, 180
Equalisation network, cascade addition, 92
Erie Autoplace, 176
Evans, C. R., 229
Exoskeletons, 147, 158
Extrapolation, 11
Eye,
  binocular vision, 192–194
  blinking, 197–198
  colour sensitivity, 194
  distribution of retinal cells, 190

Eye continued
  edge detection at retina, 215, 221–223
  fibre optic, 232
  focusing, 190–191
  following moving object, 233
  human, 219, 228
  Iitri, 232
  information capacity, 188–190
  movement, 228, 234
  muscles, 228
  robot, 127, 190–192, 194, 234, 236, 238
    automatic focusing, 236
    cleanliness, 197
  sensitivity of, 189
  single-cell, 193
  tremors, 229
  see also Retina

Failure(s),
  actual pattern of, 279
  Gaussian distribution, 284–285
  mean time between, 278, 282, 283, 287, 288
  number of, against time, 286
  overall pattern, 285
  Poisson distribution, 280–285
  probability, 274, 287
Failure rate, 281
  examples, 283
  human, 288
  unequal, 283
Failure rate curves, 279
Farm tractors, automatic control, 114
Feedback, 107, 191, 245, 251, 264, 265
  acceleration, 39
  closed loop, 33
  force, 31–33, 155, 158
    sensors, 31–33
  high-fidelity systems, 143
  position, 33–34
  tactile form, 140
  to operator, 142
  velocity, 38
  visual, 249
Feedback control, 47, 106
Feedback frequency response, 143
Feedback gain, 94
Feedback parameters, 96
Feedback stabilisation, 91, 93

296   Index

Feet, robot, 132
Ferrell, W. R., 118
Ferroresonance, 101
Fibre optic eye, 232
Fibre optics, 230, 231
Filter-rectifier circuit, 258
Filters,
  active, 259, 261
  active low-pass, 265
  mechanical, 262
  passive, 258
Fingernails, 46
Fingers, 49
  grasping movements of, 48
  pneumatically powered, 167
Fingertips, deformability of, 46
Fire protection, 195
Fisher, K. P., 202, 205, 207
Flame-sensing equipment, 195
Flanagan, J. L., 247
Fleximan, 176
Flood-lighting, 268
Flying robots, 115
Flying-spot system, 217
Force feedback, 155, 158
Force feedback sensors, 31–33
Force reflection, 140
Force transducers, 80
Force/velocity relationship, 43
Fourier transformation, 205–206
Foxall, I., 103
Frank, A. A., 152
FREDDY device, 127
Free, C. E., 221, 223
Freon, 82
Freon-12 gas, 77
Frequency analysis, 257
Frequency bandwidth, 247
  limitation of, 248
Frequency response, 23, 262
Frequency-time elements, 257
Friction, 81
Fruit-picking machines, 18
Fuel cells, 85
Fuse links, 14

Gabor information elements, 257
Gas detection, 26
Gas-sensing device, 27
Gauge factor of strain gauge, 37
Gaussian distribution of failures, 284–285

Gearbox, 59, 66
Gearing, 60, 61, 62, 66, 140
General Mills manipulator, 180, 181
'Globe' type permanent-magnet d.c. motor, 61
GM/PaR Little Ranger free-ranging vehicle, 119
Godden, A. K., 49
Goertz, R. C., 143
Golem, 9
Governor, 74
Grasping movements of fingers, 48
Gregory, R. L., 191
Grid coding, 225
Grieve, D. W., 152
Grippers, 165
Gripping device, 48
Guided weapons, 115, 149, 268
Gunn-effect diode, 239
Gyroscope, 153

Hall, J. I., 153
Hands,
  artificial, 146
  computer-controlled, 126
  magnetic or electro-magnetic, 64
  manipulator, 145
  mechanical, 165
  requirements, 46–49
  robot, 127, 139
Handwriting, use with computers, 215
Handyman master-slave manipulator, 158
Hardiman Exoskeleton, 158
Harrison, P., 260, 261
Hay, J. C., 202
Hearing,
  binaural, 252
  directional, 252
  human, 242–256
  robot, 234, 251, 257
  see also Ear
Heat dissipation, 74
Heat removal, 68
Heat sensors, 25
Heat-sinks, 69
Heating requirements, 70
Heginbotham, Professor, 175
Hendon valve, 78
Hill, D. R., 246

Hitachi robot arm (HIVIP Mk 1), 127–129, 175
Hopkins, D. G., 212, 213, 215
Hovercraft, 116
Hubel, D. H., 215
Hughes Mobile Robot (or Mobot), 119–120
Human muscle, 42
Human senses, 17, 22
Human skin, coefficient of friction of, 49
Human strength, 43
Human transfer function, 44
Human vision, 187–199; see also Eye; Vision
Humidity sensing, 28
Hydraulic accumulator, 82
Hydraulic actuation; see Actuators, hydraulic
Hydraulic drives, 181
Hydraulic fluid, leakage of, 180
Hydraulic motor, 171
Hydraulic power supply, 82, 172
Hygrometer, 27
Hysteresis effect, 52, 73

'Iitri' eye, 232
Immortality, 274
Inductive proximity, 269
Industrial robots, 163, 277, 291
Information theory, 247
Inhibition, 202, 203
  external, 204, 205
  internal, 202
Inman, V. T., 43
Inspection function, robots for, 14
Insulation, 68
Integrated circuit, OPT-5, 223
Iris control, 192

Jacobson, H., 189, 247
Jarvis, B. W., 15
Jaws, pick-up head, 166
Jindivik target aircraft, 116
Jordan, N., 289
Joshua, robot device, 8
Julesz, B., 194
Jupiter, 16

Karlin, J. E., 188
Katharometer, 27
Katys, G. P., 159
Keidel, 24
Kelley, 189
Kidd, P. A., 102
Kiessling valve, 78, 80
Klein, P. M. V., 59
Kletsky, 275
Knee joint, flexible, 154
Krab, 182
Kremer, K. H. E., 43

Ladefoged, 250
Lamb, 275
Language,
  machine recognition of, 257
  translation of, 245
Larnyx, artificial, 251
Lasers, 240
Laws of robotics, 3
Learning machines, 12, 187, 245, 258, 264
Legs,
  prosthetic, 152, 154
  robot, 132, 152, 154
  see also Walking
Lens, robot, 237
Lettvin, J. Y., 215
Licklider, J. C. R., 188
Liefer, I., 205
Lifting action, 50–52
Lift-off times, 141–142
Limbs, 139–162
Lindbom, T. H., 178, 179
Lloyd, C. J., 102
Loading and unloading devices, 166, 173, 175
Loading machines, 14, 106
Lord, M. 77
Lucas radial pintle pump, 82
Luna-16, 107
Luna-20 spacecraft, 118
Lunakhod mobile device, 107, 132–133
Lunar Rover, 133
Lunar vehicles, 159

McCarthy, 126

McCulloch, W. S., 24
Machine tools, 149, 163, 165
McKay, D. M., 230
McKenzie, D. S., 152
McKibben muscle, 77
McLeish, R. D., 83
McMillan, A., 258
McNaughton, R., 274
Madaline, 201
Magnetic circuit, 34–35
Magnetic detection methods, 268
Magnetic drum memory, 128
Magnetic pulse generation, 34–35
Magnetometers, 111
Mail sorting, 6
Manipulators, 48, 180
  bilateral control, 142
  closed-loop, 144
  control-box-operated, 145
  electrical master-slave, 144–145
  Handyman master-slave, 158
  human-controlled, 140
  mechanical form, 141
  mobile, 118–124
  powered, 142
  typical specification, 143
  remote-controlled, 117, 139–142
  speed of operation, 140, 142
  undersea, 181–183
  surface-operated, 183
  unilateral control, 142
Man-Mate industrial materials handling boom, 158
Marklew, C. M., 44
Marsh, J. F. D., 83
Mascot mobile manipulator, 120–124
Massey, R. C. G., 100
Matsushita Company, 86
Mean time between failures, 282, 283, 287, 288
Memory storage device, 239
Michie, D., 126
Microphones, 254–255
Middle ear, 243
Minataur, 143
Mine detector, 8
Miniaturisation, 68
MINIMAN, 175
MiniManip, 141
Mining, 16
MINITRAN, 175
Minsky, 126
Misalignment, 163–164
MIT projects, 126

Mk-17 program, 205
Mobile Remote Manipulative Unit (MRMU), 119
Mobile robot. *See* Robots
Mobility, 106–138
Mobot device, 183
Moiré effect, 230
Moiré pattern, 206
Moisture detection, 27
Monostable multivibrator, 265
Montgomery, S. R., 59
Moore, 274, 286
Mosher, R. S., 157
Motor, electric; *see* Electric motors
Motor theory, 245
Movement perception, 228–241
Muscles, 42–56, 77
  dynamic performance of, 42
  eye, 228
  human, 42

Nathanson, L. M., 191
National Committee for Nuclear Energy (CNEN), 120
Navigation, 39, 149
Near field control, 124–126
Neoprene seal, 76
Nervous activity, 23, 26
Nichols, G. K., 153
Nievergelt, J., 201
Nightingale, J. M., 97
Noise,
  electrical, 278
  electrical drives, 62–63
  optical, 230
Noise levels, 62–63
Nonlinear systems, 97, 145
Normalisation process, 218
Northern Ireland, 8
Notch filters, 54
Nottingham University, 175
Nuclear environment, 180
Nuclear Rocket Development Station, Nevada, 119
Numeral recognition, 263
Numerical aperture, 232

Open-loop response, 93, 94, 96
OPT-5 integrated circuit, 223
Optical illusions, 196, 238

Optical noise, 230
Optical systems, self-cleaning, 197
Orloff, 44
Outline enhancement by superimposition, 217
Oxygen analyser, 27

Paget, R., 242
Pain sense, 22
Papert, 126
PAR3000 manipulator, 145
Parker Hanifan hydraulic cylinder, 169
Passenger-carrying vehicles, 7
Pattern-matching gates, 209
Pattern recognition, visual, 187, 209
Paul, J. P., 152
Paul, R., 175
Perceptron, 187, 201, 202
Performance, 91
  requirements, 91–92
  specifications, 91
Phase-advance in receptor circuits, 23
Phase advance network, 92, 93
Phonemes, 267
Photo-cells, 206–223, 231, 237
Photo-detector, 193, 195
Photo-diodes, 223
Photo-electric equipment, 197
Photo-multiplier tube, 217
Photo-transistors, 223
Picturephone system, 192
Pierce, J. R., 188
Piezo-electric strain detectors, 37
Piezo-electric vibrator, 263
Piezo-resistive effect, 37
Pilling, H. W. C., 14
Pilot, automatic, 116
Pistecky, P. V., 81
Planobot, 166–67
Pneumatic actuation; see Actuators, pneumatic
Pneumatic valve, 81
Poisson distribution of failures, 280–285
Poisson probability series, 281
Population trends, 11
Position determination, 109
Position feedback, 33–34
Position measurement, 33, 123
Position sensors, 33, 106
Positioning, 228

Positioning table, 175
Posture, 149
Potentiometers, 167, 173, 176, 237
Power requirements, 86
Power sources, 72–90; *see also* Batteries
Power supply,
  electrical, 83–86
  hydraulic, 82, 169, 172
  see also Batteries
Pressure gauges, 32
Pressure sensing, 22
Pressure/volume pneumatic actuation curve, 75
Printed circuit, 61
Probability density, 284
Process control equipment, 279
Programmed and Remote Systems Corporation, 141
Prosthetic devices, 31, 32, 42, 44, 48, 52–53, 62, 73, 77, 82, 85, 97, 146–147, 152–154, 175
Proximity, inductive, 269
Proximity detector,
  capacitive, 270
  radio-frequency, 268
  using bridge technique, 271
Proximity switch, 164
Pulse amplitude modulation, 78
Pulse code modulation, 209
Pulse frequency modulation, 78
Pulse pattern detectors, 209
Pulse width modulation, 78
Pump, Lucas radial pintle, 82

Quadratic equations, 97
Queen Bee target aircraft, 116
Queen Wasp target aircraft, 116

Racer programme, 275–276
Radar,
  Doppler methods for tractor speed measurement, 115
  microwave devices, 268
  miniature, 239–240
Radiation, 269
Radio communication, 125
Radio control, 125
  aircraft, 116
  target speed-boats, 116

300 Index

Radio-frequency interference, 125
Radio-frequency proximity detector, 268
Rakic hand, 48, 49, 80
Ralston, H. J., 43
Rancho Los Amigos orthotics, 120
'Rand' tablet, 215
Receptor nerve cells,
  response of, 23
  transient response, 25
Reddy, D. R., 246
Redundancy in character recognition, 215
Redundant parts, 287
Rees, M.G., 236
Reflex activity, 25–26
Regulator systems, stabilisation of, 92
Reliability, 6, 7, 9, 101, 274–287
  in terms of mean time between failures, 282, 283
  method of improving, 287
Reliability probability, 287
Repair, self, 275, 278
Repair time, 289
Reproducibility, 276
Research, 291
Resistance, variation with temperature, 30
Retina,
  artificial, 219, 228, 231
  artificial scanning, 206, 209
  counting, 219–221
  edge-detecting, 215, 221–223
  robot, 230, 232
  see also Eye
Ring, N. D., 49
Rivet machine, 126
Road Research Laboratory, 114
Roberts, L. G., 225
Roberts, T. D. M., 154
Robertson, A., 100
Robot limitations, 4
Robot Systeme, 151
Robotics, laws of, 3
Robots,
  and humans, comparative economics, 177
  climbing, 149, 154, 155
  copying, 8, 163
  domestic, 3, 4, 5, 63, 107, 180, 181, 291, 292
  effectiveness, 277
  existing devices, 5

Robots and Humans *continued*
  fictional, 2
  flying, 115
  form of, 2
  future of, 291
  general-purpose, 1, 9, 10, 11, 12, 13, 32, 33, 47, 48, 57, 68, 69, 154, 248, 252, 253, 291, 292
  guided, 107
  humanoid, 1, 2, 228
  industrial, 163, 277, 291
  infra-red sensitive, 195
  man's fear of, 2
  minimal requirements, 3–4
  mobile, 106–138, 154, 181, 240, 247, 254, 278, 291
    'Astor', 102–104
    computer-controlled, 126–130
    investigation of problems, 101–204
    legs, wheels or tracks, 131
    near field control, 124–126
    restrictions, 107
    see also Mobility
  rail restricted train, 107
  self-repairing, 275, 278
  self-reproducing, 275
  space; see Space
  special-purpose, 12, 194–195, 248, 252, 291
  ultra-violet-sensitive, 195
  vehicle-driving, 108
  walking, 82, 132, 153, 155
  see also under specific types and applications
Robotug, 9, 108–110
  advantages of, 113
  safety measures, 112–114
Rogers, G. L. 205

Safety measures,
  Robotug system, 112–114
  with force reflection, 140
Scanning artificial retina, 206–209
Scanning system,
  computer, edge detection in, 224
  contrast enhancement in, 216
Schottky diode, 223
Schroeder, 219
Sea-bed, sampling device, 182
Seal, Neoprene, 76
Selection, 275

Self-Propelled Anthropomorphic Manipulator (or Sam), 120
Self-repairing robots, 275, 278
Self-repairing system, 275
Semicircular canal system of inner ear, 148, 153
Semiconductor, diode, 31
Semiconductors, 15, 28, 29, 36, 37
Sense organs, touch, 271
Senses, 17–18, 22–41
Sensing devices, 17
Sensors,
  balance, 148, 149
  electromagnetic, 149
Servo amplifier, 124
Servo control, 145
Servo-motor, 123
  step, 67–68
Servo system, 121, 144, 170, 176, 191, 236
  'bang-bang' or 'on-off' type, 72
Shadow mask technique, 202
Shape discrimination, 32
Sheridan, T. B., 118
Shielding, 180
Shift register, 209, 210
Shunt resistance, 30
Sideman, 176
Signal detection, self-produced signals, 267
Simplicity, 277
Simpltran, 173
Single equivalent formant, 248
Size requirements, 68
Skin, 49
  human, coefficient of friction, 49
Smell sense, 26–27
Solenoids, 80
  actuators, 63–65
  rotary, 65
Soutter, L. D. L., 258
Space applications, 59, 60, 70, 107, 115, 116–118, 132–133, 159, 180, 240, 278
Space manipulator, 117
Speech,
  information content, 247
  machine recognition, 257
  nature of, 246
  robot, 266
  structure of spoken words, 249
Speech recognition, 243, 244–246, 249, 250, 251
Speech sounds, components of, 248

Speech waveform, 246
Speed-boats, radio-controlled, 116
Speed control, 74
  continuously variable, 98
  Doppler methods for tractor, 115
Stabilisation, 91
  feedback, 91, 93
  of regulator systems, 92
  simplified, 91–96
  extension of, 96–98
Stability, 91–105
  closed-loop, 93
Stanford hand-eye system, 224
Stanford University, 126
  Artificial Intelligence project, 129–130, 175
Statistical delay, 44
Statistical tests, 284
Steeper locking device, 47
Steering systems, 114, 115
Stephens, N. W. F., 205
Stepping device, 66
Step-servo motors, 67–68
Steromotor, 66–67
Stiction, 58, 80
Stock control system, 150
Storekeeper, 150
Strain detectors, piezo-electric, 37
Strain gauge, 31, 32, 35–37, 78, 80, 81
  gauge factor of, 37
Submersible, 183
Sulphur hexafluoride, 82
Superimposition, outline enhancement by, 217
Supervisory function, 278
Surveyor moon flight, 16, 117
Switching devices, implanted, 272
Szabo, M., 169, 178

Tachometer brush ripple, 39
Tachometer generators, 38
Taguchi Gas Sensor, 27
Target aircraft, 116
Tarp (Test and Repair Processor), 278
Taste sense, 27
Taylor, A., 45
Taylor, W. K., 202
Teaching device, 188
'Telechiric', use of term, 126
'Telechiric' devices, 16
Telemotive system, 125

Telenaute, 183
Television,
  colour, 194, 237
  three-dimensional, 193
Television cameras, 117, 120, 121, 126, 127, 129, 143, 182, 197, 216, 217, 224, 235–238
Temperature compensation, 36
Temperature control using thermistors, 29–31
Temperature regulation, 28–29
Temperature stabilising system, 70
Temperature variation with resistance, 30
Terresearch, 182
Test function, robots for, 14
Testability, 277
'Therbligs', 163
Thermilinear components, 31
Thermistors, 29–31, 36
Thermocouple, 29
Thermopile, 29
Thring, Professor M. W., 3, 4, 5, 49, 126, 132, 150, 154, 164
Thumb movement, 48
Thyristor, 98, 259, 262
Time constant, 58, 92, 95, 96, 97, 145
Time delay, 58
Time delay relay, testing, 15
Time sense, 18
Todd, R. W., 97
Tomovic, R., 48
Torque motor, 66
'Tortoise' machines, 101
Touch sense, 22, 32–33, 271
TRACERS, 277
Tracks, 131
Tractors, farm, automatic control, 114
Trade unions, 10
Trains, underground, 7
Trallfa robot, 176
Transducer pads, 32
Transfer function, 44
  reshaping of, 93
Transfer machines, 163
Transferobot transporter, 151
Transient response, 97
Transistor, 31, 53
Transiva, 176
Translation of languages, 245
Transrobot, 151
Tremor, 58
Trieste, 181

Tyres, 131

Ultrasonic communication, 125
Ultrasonic 'eye' 268
Ultrasonic radiation, 268
Ultra-violet-sensitive robot, 195
Underground trains, 7
Underwater applications, 142, 146, 180, 181–183
Underwater vision, 195
Unemployment and redundancy, 8–9, 12–14, 291
Unimate, 164, 167–171
  control system, 168
  economics of, 178
  memory capacity, 170
Unimo, 183
Uniselector, 44, 66, 207, 208
United States, 291
Utilisation, optimum, 289

Vehicle-driving robots, 108
Vehicles,
  battery electric, 107
  passenger-carrying, 7
  urban transit, 110–112
Velocity feedback, 38
Velodyne, 238
Versatran, 48, 164, 171–175
  continuous-path, 174
  economics of, 177–178
  point-to-point controller, 172
Vibrotactile devices, 24
Vicens, P. J., 246
Victoria Line, 7
Vietnam, 7
Vision,
  binocular, 192–194, 237
  human, 187–199
  persistence effect, 238
  robot, 212–227
  see also Eye
Visual pattern recognition, 187, 209
Vladievskii, A. P., 6
Vocoder, 243, 266, 267
  for use in robots, 261
  with active filters, 259, 261
  with mechanical filters, 262
  with passive filters, 258

Voder, 266
von Kempelen, Wolfgang, 242

Walking,
  comparison of human and robot, 151–154
  on sloping surfaces, 154
Walking machine, 82, 153
  quadruped, 155–157
Walking robots, 132
Walter, W. G., 102, 235, 288
Ward-Leonard system, 100
Warner, M. G. R., 114
Warner, R. M., 252
Water content of body, 29
Water vapour detection, 27
Waveform, speech, 246
Wear problems, 54

Weber-Fechner law, 18, 191
Wee, W. G., 201
Weight, limitations, 59
Weight sensors, 32
Weimer, P. K., 223
Wheel as motive source, 131
Wiesel, T. N., 215
Williams, P. S., 219
Witt, D. C., 152, 153
Worm gear, 60, 61, 72, 146
Wright, R. H., 27
Wrist action, 47

Yellow Springs Company, 31

Zener diodes, 31